Image Processing
of Geological Data

COMPUTER METHODS IN THE GEOSCIENCES

Daniel F. Merriam, Series Editor

Computer Applications in Petroleum Geology
Joseph E. Robinson
Graphic Display of Two- and Three-Dimensional Markov
  Computer Models in Geology
Cunshan Lin and John W. Harbaugh
Image Processing of Geological Data
Andrea G. Fabbri

IMAGE PROCESSING OF GEOLOGICAL DATA

ANDREA G. FABBRI, Geological Survey of Canada

 VAN NOSTRAND REINHOLD COMPANY

To Finella

Manufactured in the United States of America.

Published by Van Nostrand Reinhold Company Inc.
135 West 50th Street
New York, New York 10020

Van Nostrand Reinhold Company Limited
Molly Millars Lane
Wokingham, Berkshire RG11 2PY, England

Van Nostrand Reinhold
480 Latrobe Street
Melbourne, Victoria 3000, Australia

Macmillan of Canada
Division of Gage Publishing Limited
164 Commander Boulevard
Agincourt, Ontario MIS 3C7, Canada

15 14 13 12 11 10 9 8 7 6 5 4 3 2 1

Library of Congress Cataloging in Publication Data
Fabbri, Andrea G.
  Image processing of geological data.
  (Computer methods in the geosciences ; 3)
  Includes index.
  1. Geology — Data processing.   2. Image processing.
I. Title.   II. Series.
QE48.8.F29   1984   550'.2854   84-3729
ISBN 0-442-22536-9

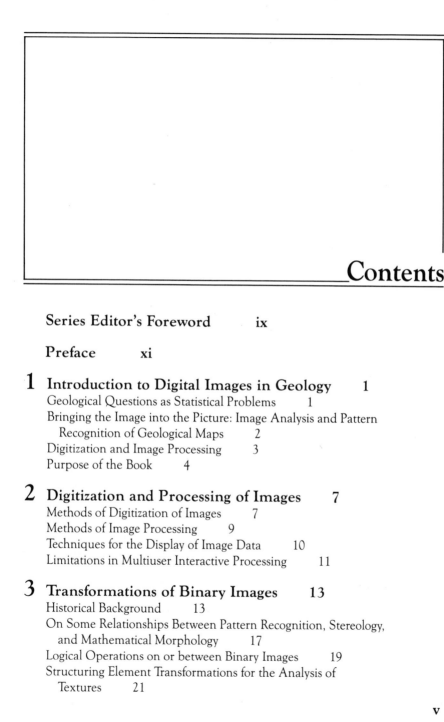

Contents

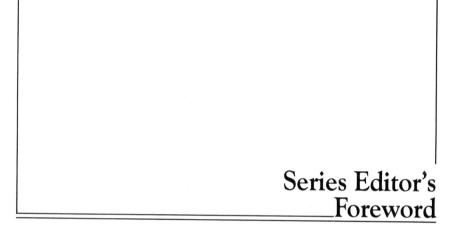

Series Editor's Foreword

Image analysis is not new to geologists because they have been doing it visually for many years. The digitization, processing, and interpretation quantitatively, however, is new. Image analysis and pattern recognition of geological maps is the basis of much geology whether done qualitatively or quantitatively.

Quantitative aspects of geology go back to the roots of the science, but the application of computers is relatively new and has served as a catalyst in quantifying the subject. The past 25 years has seen a revolution in geology with the development of plate tectonics, geomathematics, planetology, seismic stratigraphy, emphasis on the three Es—energy, environment, and ecology—and much more. The technological advance accompanying this revolution was the rapid development, acceptance, and adoption of computer methods. The ability to acquire, manipulate, and analyze massive amounts of data is extremely important. Image analysis is no exception—without the computer, all that is reported in this book would be impossible.

The development of image making, then, is dependent on the development of new technology, both hardware and software. The widespread use of small, large capacity, fast machines coupled with available computer programs such as GIAPP has made image processing feasible and economical. Thus, it makes little difference if the researcher is working with LANDSAT imagery or rock thin sections, the process is the same.

Dr. Fabbri has put together for the geologist a book on how "to study geological maps and microscopic images of thin sections of rocks by methods of image processing." Included in the discussion is the philosophical approach, the software, and practical applications. With help of Dr. Fabbri's book, researchers and graduate students should be able to apply the procedures similarly to their problems and resolve them. Followers of many disciplines other than geology should find this presentation of value and interest.

The Computer Methods in the Geosciences series is designed to fill a specific need—bring to geologists computer-oriented techniques in a manner and language that is understandable and usable. Dr. Fabbri's contribution fulfills the requirements well. He is a practicing geologist using the methods described in the book to solve real problems. His many years of experience in the field makes him one of the leaders in geologic pattern recognition.

This book is the second in the series—the first being *Computer Applications in Petroleum Geology* by Joseph E. Robinson. Others are in press or in preparation. The series is open ended so that topics can be added as needed or as new developments take place, and each contribution is designed to be self-contained. At present, many exciting possibilities are unfolding in computer methods and applications, the results of which are being made available through this series.

D. F. MERRIAM

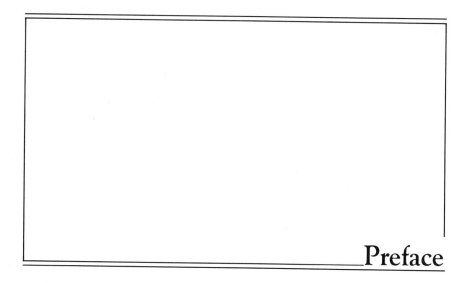

Preface

This book proposes to study geological maps and microscopic images of thin sections of rocks by methods of image processing. Such methods, which include pattern recognition and image analysis, are used for extracting information from two-dimensionally distributed data in computer form (digital images). To support the claim made that image processing and pattern recognition can be useful in geology, an interactive package of FORTRAN programs was developed for geological applications in the areas of statistical mineral-resource estimation, stereology, and image analysis by mathematical morphology.

The book spans three different aspects of geological image processing: the philosophical approach in geology, the programming of image-processing software, and practical applications of various complexity, including one pattern-recognition experiment. The study is aimed at applications of the dimension and character of conventional real-life geological studies, and also at relatively large digital images of about 1 million picture elements.

Many geological problems (and indeed, problems in many other fields) are well suited for solution through image processing, especially when the data are obtained at discrete locations and can be represented in map form or in the form of a digitized photograph and other imagery. Throughout the book, the visual aspect is stressed by the many illustrations. The subject lies at the triple point of geology, computer science, and geometric probability; the aim is to familiarize the geologist with the image-processing approach by way of example.

The book is intended for researchers and graduate students in the geosciences and related disciplines that use maps and microscopes: agriculture, forestry, environmental sciences, geography, land-use planning, remote sensing, petrology, mineralogy, and material sciences. Researchers examining biomedical thin sections will also find techniques applicable to their problems.

A general statement on geological images and how they can be useful is made first in Chapter 1. Chapter 2 introduces the digitization and processing of images, and Chapter 3 describes transformations of binary images. Some simple examples of binary geological images obtained from a map pattern and from a microscope image of a thin section are discussed in Chapters 4 and 5. Next, Chapters 6, 7, and 8 describe a case-history application to the study of regional mineral resources. In that application, large images are used to show how such images must be kept in registration with one another. The description of the experiments is preceded by a detailed account of the required digitization and preprocessing steps in Chapter 6. A geological data base constructed for a study area in northwestern Manitoba and described in Chapter 7, is used in Chapter 8 to introduce the reader to the quantitative characterization of geological and ancillary map patterns and to establish their relationship to the distribution of mineral occurences in the area. Chapter 9 describes the scanning of line drawings from thin sections of rocks and the preprocessing related to this type of digitization by a flying spot scanner. Two applications are made to a previously studied granulite in search for a pattern (Chapter 10) and to an amphibolite (Chapter 11) for comparing the performance of an image analyzer (hardware-based approach) with that of the software pro-grammed on a minicomputer (software-based approach). In Chapter 12, studies are reviewed of rock textures for the analysis of microscopic thin and polished sections. Chapter 13, "Towards Pattern Recognition," describes a first experiment of truly automatic processing. Finally, Chapter 14 is a summary epilogue.

ACKNOWLEDGEMENTS

D. F. Merriam, of Wichita State University, suggested the preparation of this manuscript from the original version of my Ph.D. dissertation at the University of Ottawa.

F. P. Agterberg, G. F. Bonham-Carter, C. F. Chung, Tonis Kasvand, Stefano Levialdi, Jacques Masonuave, William Petruk, and Giorgio Ranalli critically reviewed parts of the manuscript and made many helpful suggestions.

The Geological Survey of Canada and the Electrical Engineering Division of the National Research Council of Canada generously supported the research leading to the completion of the book.

ANDREA G. FABBRI

CHAPTER 1

Introduction to Digital Images in Geology

GEOLOGICAL QUESTIONS AS STATISTICAL PROBLEMS

In the natural sciences, students are confronted with a real world of phenomena that may seem continuous in time and physical properties. Their task is to collect observations over an area, a volume, or a time interval. The planning of these observations may be random at first, or may follow guidelines learned previously elsewhere. With the accumulation of observations, systematic patterns emerge, which seem to correspond to phenomena observed in different places. Such patterns provide the tools for describing the collection of observations in a comprehensible and reusable manner.

Because no natural phenomenon can be described practically in its entirety, its classification and the measurement of its properties are seldom clearly definable tasks. In general, the student resorts to assigning a certain degree of trust to each observation, and by comparing several observations he is likely to increase his confidence about phenomena occurring at smaller scales. Thus the student progresses slowly until he feels comfortable making statistical statements. These statements represent only in part the properties of the phenomena studied and must be viewed in terms of the observation plan.

Some geological questions have been formulated as quantitative models to which statistical tests can be applied. Many problems have become both geological and statistical: a new discipline termed "mathematical geology" was established about 15 years ago (Vistelius, 1969). Several fundamental statistical questions in geology are older than that, however, as became apparent in the work of the

1

mineralogist M. A. Delesse, who, as early as 1848, indicated that the relative volumes of minerals in rocks could be measured from random sections. For a recent overview of statistical applications and their historical development in the study of spatial patterns in the earth sciences, see Agterberg (1978) and Merriam (1981).

Modal analysis, quantitative stratigraphic correlation, analysis of vectorial data, trend-surface analysis, statistical simulation, numerical classification, and mathematical morphology are just some of the subjects that are becoming integral parts of geology, in almost any geological specialty, whenever practical applications are required. Pattern-recognition techniques are powerful tools in many geological applications in which spatial patterns play an important role.

BRINGING THE IMAGE INTO THE PICTURE: IMAGE ANALYSIS AND PATTERN RECOGNITION OF GEOLOGICAL MAPS

A great many geological studies cover a particular area of given size, which can be sampled at different levels of detail. Time and financial constraints, as well as other logistic limitations, tend to limit drastically the level of detail in any study. For this reason the statistical inference of a study will be limited by the scale that has been chosen. Once the scale and the consequent sampling intervals have been selected, the geological observations can be represented in map form, where the different map units subdivide all representable features and portray spatial attributes—size, distribution, and orientation—and the interrelationship between the various features. Frequently, all observations cannot be represented on a single map, so several maps—each covering a different theme, such as lithology, structure, economic geology, geophysics, geochemistry, and engineering geology—must be prepared.

Each unit on a map has its own set of geometrical properties, some of which are important for describing particular geological characteristics. It is desirable to capture these properties systematically, so that statistical estimates of the attributes of the map units can be computed easily.

Geological maps contain a great amount of information. Not all of it is used in most geological studies, but relevant features are considered. The selection of these features is seldom systematic and dictated by quantitative criteria. The methods used in this book attempt to bring geological images "into the picture," thereby extending computations and interpretations to the geometric attributes of geological representations.

Geological maps can be transformed into digital images for picture processing. Their attributes can be measured automatically so that it becomes possible to recognize patterns of relationships that may contribute to explain geological phenomena in quantitative terms. Images of microscopic rock material also can be transformed into digital images and considered as micromaps covering small geological areas.

2

DIGITIZATION AND IMAGE PROCESSING

Picture or *image processing, pattern recognition,* and *scene analysis* are some of the terms used to describe methods of extracting information from (usually) two-dimensionally distributed data. Such data are called *digital pictures,* or *images.* Our own visual system — the eyes and brain — is extremely well suited for qualitative interpretations of pictures. Our innate ability to carry out quantitative measurements "at a glance," however, is extremely coarse. For a computer the situation is the opposite: accurate measurements and extensive computations are easily programmable whereas it may be difficult or even impossible to program the computer to identify what to measure. For this reason, "interactive methods" can be successful, in which data processing is simply shared between the human and the computer, each doing what is "easiest."

The number of computerized picture-processing methods is so large, and the variety so wide, that only the main approaches have been summarized (Rosenfeld and Kak, 1976). Typically the entire image, rather than just selected areas, is processed for the information sought.

Computers analyze pictures in digitized form: the computer memory stores the digital representation of pictures as arrays of numbers. A gray-level, or density, value is associated with, and in point-to-point correspondence with, each small area of the original picture. These numerical arrays are used to measure the different parameters of the patterns in the picture. A description of a picture involves the enumeration of properties of the picture or of its parts, and of relationships among the parts. For example, geometric properties do not depend on the gray levels, but only on the sets of picture points, or *pixels,* that belong to given parts of the picture.

Pictures can be digitized in several ways. Two usual methods are scanning by optical or mechanical devices, and digitizing of lines by x-y digitizers. Both techniques were applied in the preparation of this book. A flying spot scanner was used to scan 35-mm transparencies of black-and-white tracings of map-unit boundaries, and an image analyzer was used to study black-and-white transparencies of maps; a graphic tablet was used to digitize boundaries of crystal grains from the tracings of rock textures and boundaries from map patterns.

Geological maps can be computer processed on a general-purpose computer with a FORTRAN compiler. The program package used in this book is based on a model-interactive image-processing system programmed by Dr. T. Kasvand at the Electrical Engineering Division of the National Research Council of Canada. Designed especially for the analysis of geological images, the system is called the Geological Image Analysis Program Package (GIAPP) for estimating geometrical probabilities. GIAPP analyzes maps or microscopic images in terms of their geometrical attributes, which can be estimated statistically. For this reason, methods of mathematical morphology, geometrical probability, and image processing were considered in the design. The package consists of file handling, interactive conversation, picture display, image digitization, and editing routines, in addition to

3

algorithms for measuring parameters of black-and-white images by methods from different fields of study. See Appendix A for an overview of the computer system used and Appendix B for a summary description of GIAPP.

PURPOSE OF THE BOOK

This book aims to provide geologists with new techniques for analyzing geological (and ancillary) maps. The techniques capture the geometrical configuration of features and offer the choice of different sampling schemes according to model requirements. The approach is based on (*a*) the quantification of images by a software-dependent flying spot scanner (which allows control of geometrical configuration of the scanning pattern), (*b*) the use of a graphic tablet, and (*c*) computer programming for the interactive analysis of digital images.

Computer algorithms were developed with the aid of techniques and statistical methods used in different disciplines. This made the software more general and flexible because the fields overlap to a great extent in terms of the type of problems being studied but not in terms of the methods applied.

Many instruments are available for image analysis. They are built according to a modular concept and have consisted until recently mostly of hardware. They are expensive, and their acquisition and use can be justified with production work. It is likely that soon, with the increased availability of cheaper computer memory and microprocessors, the manufacturers will add programmable computers to these instruments, making it possible to use more complex and satisfactory image-processing methods. For research, where experiments are performed only on limited numbers of images, computer processing, even with a small computer, offers more flexibility of approach and lighter capital costs if a computer is already available and if the software is provided. This is where GIAPP is important: it is an interactive image-processing package that in part duplicates what those instruments can do and in part expands their capabilities. The logic of the various algorithms in the package is documented fully, and it complies with the theories developed in mathematical morphology for texture analysis. In addition, the GIAPP package contains several image-processing algorithms that expand the theory and applications of textural studies.

Much geological sampling of microscopic and macroscopic images may be bound by rigid, preset schemes; for example, point counting for modal analysis or for map data quantification. The measurements made in this manner are time-consuming, such that repeated sampling is seldom possible. This book describes how images can be input into computers to make more satisfactory measurements on them.

Because the problem of improving and accelerating input preparation for images that are produced directly by a microscope is complex and essentially material-dependent, it is not considered here. The basic target of the book, then, is the development through computer programming of a methodology that is related directly to important geological applications. Applications of different types have

4

been chosen, to emphasize the adaptive character of software development and the applicability of methods of image analysis. The methods have broader application outside the purely geological domain.

The following geological applications are considered here: geometrical probabilities associated with geological and geophysical maps, textures in drafts from thin sections of metamorphic rocks, feature extraction from micrographs of alpha tracks, and the extension of theory and applications to general texture problems. The results of these approaches usually will be expressed in graphic form to aid the communication of the various steps in processing images by computer. Because it was important to experience concretely the consequences of the suggested approach with applications of the dimensions and the character of conventional geological studies, considerably large (1000 pixels × 1000 pixels) images were analyzed at some depth for either geological map patterns or microscopic images from thin sections of rocks.

REFERENCES

Agterberg, F. P., 1978, Analysis of Spatial Patterns in the Earth Science, in *Geomathematics, Past, Present, and Prospects*, D. F. Merriam, ed., Syracuse University Geology Contribution 5, pp. 7-18.

Delesse, M. A., 1848, Procédé Mécanique pour déterminer la Composition des Roches, *Ann. Mines* **13:**379-388.

Merriam, D. F., 1981, The Roots of Quantitative Geology, in *Down-to-Earth Statistics: Solutions Looking for Problems*, D. F. Merriam, ed., Syracuse University Geology Contribution 8, pp. 1-15.

Rosenfeld, A., and A. C. Kak, 1976, *Digital Picture Processing*, Academic Press, New York, 457p.

Vistelius, A. B., 1969, Preface, *J. Math. Geol.* **1:**1-2.

CHAPTER 2

Digitization and Processing of Images

Digital images are best treated by computers interactively. Processing is done in steps in which an image is transformed and then displayed to allow the operator to decide which further transformations may be required. The steps depend on the material analyzed and represented in the picture: human interaction is needed until enough is known about the image for a process to be designed for automatic pattern recognition.

First a picture is digitized into a digital image, which undergoes various types of transformations for image enhancement and feature extraction, termed *preprocessing*. Then the image is processed for recognition and classification.

METHODS OF DIGITIZATION OF IMAGES

Two-dimensional pictures can be quantified through digitization, which generally is termed *quantization*. The procedure consists of assigning numerical values to discrete points of given coordinates. The values are proportional to the "gray level" (optical density, transmittance, reflectance, etc.) of the elementary unit of observation. In digital processing the picture samples must be quantized; thus the range of gray-level values in the samples must be divided into discrete intervals and all the values within an interval must be represented by a single level. Generally the samples are obtained by using an array of points.

Optical digitization is done by instruments known as scanners (e.g., micro-densitometers, flying spot scanners, image analyzers), which break the image into

7

many units, each representing the tone or density in the immediate neighborhood of a point in a regular array. This array (pixel, or point, configuration in the structure of image data), termed a *raster*, is either square or hexagonal.

A square raster (Fig. 2.1A) is formed by pixels spaced so that they occupy the corners of a regular square grid. The distance between adjacent pixels in the horizontal or vertical direction is 1, whereas in the diagonal direction it is $\sqrt{2}$. In the hexagonal raster, shown in Figure 2.1B, the pixels occupy the corners of a network of equilateral triangles. The distance between adjacent pixels is always 1; it can be measured along the three directions of the network. In this configuration the distance between adjacent rows is $\sqrt{3}/2$.

If the raster is regular, it is not necessary to record geometrical addresses because they are implicit in the position of each observation in the series. The tone or density (gray level) of each raster point must be expressed as one of a limited number of digital characters. For black-and white-material, it is sufficient that each observation show only that the point was in black and not in white, that is, in the binary 1 and not in the 0, which are convenient values to symbolize those two gray levels.

For more complex structures, a separate character is required for each of the gray levels. If these are not spaced uniformly in the available tone range, several extra characters are required. Such a multiple-level record will be treated as the source of several binary images, each representing one of the "phases" to be measured. *Phase* here refers to the set of points that includes all particles of a single type (here gray-level range, or color). This concept is important because it allows the analysis of complex structures to be treated as a set of binary problems. This treatment allows a computer equipped with Boolean operators to process an image at a faster rate than when each image point is processed as a numerical character.

A flying spot scanner performs raster scanning, but it is a sophisticated and expensive piece of equipment. There are many alternative instruments, however.

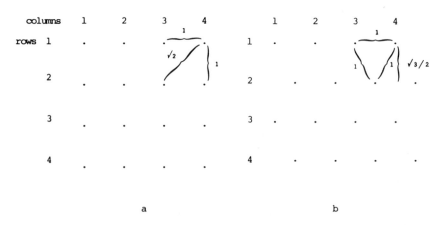

Figure 2.1. Diagrammatic structure of square (A) and hexagonal (B) raster configuration of pixels.

The drum scanner and the flatbed scanner are two widely used raster input devices. In the drum scanner a picture is mounted on a cylinder, which spins on its axis. In the flatbed scanner a picture is mounted on a flat-moving platform. A beam of light is reflected off or transmitted through the surface of the picture into a photocell.

METHODS OF IMAGE PROCESSING

Computer methods for image processing in general assume that the data to be processed have a particular arrangement. To clarify this concept, it may be useful to consider image processing as opposite to computer graphics. According to the standard practice, image processing refers to operations that transform images into other images to obtain information about them and about the objects in them; image recognition is the mapping of images into nonimage description. Because problems depend on the content of the image, we cannot decide in advance what question to ask the computer about the image, or which measurements to make. Images are stored and accessed directly as large matrices of values, and computer programs do understand neither the images nor their parts. It is often said in image processing that images are stored "explicitly."

In contrast, computer graphics is concerned primarily with computer synthesis and manipulation of images that are specified by descriptions. The fundamental parts or objects belonging to the images must be known beforehand. A carefully designed and efficient database system is needed to manipulate pictures. Images are stored implicitly, that is, synthesized into particular descriptions understood by the computer programs. The information, therefore, is coded and sparse. The amount of memory used is generally small, but the data must be arranged so that the needed information can be found without much searching and computation. Some fundamental parts of databases for computer graphics are coded as edges, polygons, and various pointers (to connect the parts into objects) to these, generally structured in particular sequences. Such sequences are problems-specific to be most efficient computationally.

Picture processing, or image processing, is used for the following reasons (which also constitute the basic historical requirements for it): (a) picture digitization and coding, (b) image enhancement and restoration, and (c) image segmentation and description or pattern recognition. For these purposes, images are stored as regular arrays or matrices of values. Each value corresponds to an individual picture element or pixel, which in turn is in point-to-point correspondence with a point or, more precisely, with a small area in the original input material. Because these arrays are usually large—for example, $1000 \times 1000 = 1$ million pixels—images are stored on magnetic tapes or disks rather than directly in the memory of the computer.

In the system described in this book, image data are represented as matrices of integer or one-bit binary numbers of up to 1024×1024 elements. They are stored row by row in a sequential manner. Each row of picture data represents one logical record.

Gray-level images are digitized so that each computer word stores the gray-level value (generally an integer number) at a given position. A binary image has only two values of gray level, 1 and 0, which can be considered as black and white, respectively. Binary 0-1 data are compressed so that each 0-1 pixel corresponds an OFF-ON bit status. This is done to economize in memory storage and to expedite input/output (I/O) and some types of computations, as we will see later.

Image processing by a general-purpose time-shared computer can produce the desired results satisfactorily with a relatively modest programming effort. Such computers generally are available everywhere, but the processing is bound to take considerable amounts of computer time. The general-purpose computer is thus ideal as a research tool, but it may not be adequate for the routine production of results.

In processing binary images, one word in the computer now can contain many picture elements, and in a modern computer all the bits within a computer word are processed in parallel when at least (Boolean) logical operations are involved; a degree of parallelism is obtained that is proportional to the word length within the computer. Thus certain binary operations can be speeded up considerably.

Special "hard-wired" image-processing devices have been on the market for some time (Hougardy, 1976). These devices, generally termed *image analyzers*, are fast, but the operation repertoire is very limited. Due to the increased importance of picture processing, "pipelined" and "parallel" processing computers are becoming available. These also can be programmed in a manner analogous to the ordinary general-purpose computers (Preston et al., 1979). Such computers, however, are expensive.

The approach followed in this book is directed toward a poorman's solution in a time-shared computer: a usual situation in which we are paying in computing time and not in hardware. The richman's solution is a special-purpose computer. This is a problem that the image-processing community is actively trying to solve. Soon special image-processing "chips" (integrated circuits) will be available that will work in a so-called pipeline fashion at approximately television speeds. It seems likely that geology—as well as other disciplines such as remote sensing and signal processing,—will take increasingly more advantage of the progress in image-processing technology.

TECHNIQUES FOR THE DISPLAY OF IMAGE DATA

Many aspects of image processing are related clearly to our perception of pictures. The results of the transformations computed can be appreciated best in pictorial form as displays obtainable on various display devices. On a television screen we can have a "refreshed" type of display: the image can be stored on image buffers or frame buffers, and then displayed continuously with many different gray levels (and also colors). We can use a memory-type cathode-ray tube (CRT) device, where the storage of the image occurs on the phosphorous surface: here only three or four gray levels can be displayed for each single dot (pixel).

Various types of printers can print displays on paper. A line printer can overprint characters to produce a few different gray levels. Plotters and dot printers are much more powerful. Plotters, either drum or flatbed, can print using several color ribbons. Different picture parts can be represented by contours, graphic patterns, various characters, or drafted symbols. In dot printers each pixel is represented by a square matrix of dots. For example, by deleting dots, ten different gray-level effects could be obtained from a single color of dots, generally black, in a 3×3 matrix. Ink-jet plotters, have three guns that fire droplets of fast-drying ink of the three basic colors (yellow, magenta, and cyan) onto a special absorbing paper. The colors are overprinted in three passes, and the end results are pictures of different colors and intensities, similar to what is done in color television.

LIMITATIONS IN MULTIUSER INTERACTIVE PROCESSING

Many display devices can be used with a dedicated (single-user) computer system, such as GIAPP (see Appendix A). When only a graphic terminal is available, and when it is connected to a time-shared computer, there are some limitations on display and interaction. The graphic terminal screen and its keyboard have to be used in turn for both the interaction and the display of pictures. Thus it becomes necessary to use a permanent recording device such as a camera or, better, a printer or hard-copy unit that, whenever needed, will photographically copy what is displayed on the screen for documentation of the processing results and the sequence of the various computational steps. The latter alternative is exemplified in Appendix D.

The use of magnetic tapes may not be permitted in an interactive time-shared environment with many users; therefore only disk space may be available. Finally, because the memory space allowed in the interactive mode may be more limited than in batch mode, programs must be tailored to the allowed memory, that is, structured for minimum space requirements into overlays or into segments. Running speeds also will have to be minimal. Particularly, a relatively fast rate of transmission for the line connecting the graphic terminal to the computer is to be preferred for the display of images, for example, 4800 or 9600 bauds (480 or 960 characters per second). Failure to obtain these speeds will limit the efficiency of image display in the multiuser-interactive environment.

REFERENCES

Hougardy, H. P., 1976, Automatic Image Analysing Instruments Today, Proc. 4th Int. Congr. Stereol., Gaithersburg, Maryland, Sept. 4-9, 1975, *Nat. Bur. Stand. Spec. Pub. 431*, pp. 141-148.

Preston, K., Jr., M. J. B. Duff, S. Levialdi, P. E. Norgren, and J-i. Toriwaki, 1979, Basics of Cellular Logic with Some Applications in Medical Image Processing, *IEEE Proc.* **67**: 826-858.

CHAPTER 3

Transformations of Binary Images

Experiments on simple images are described in this chapter. They exemplify the concepts developed further on more complex practical applications on materials of geological nature. Attention is first given to the early approaches to the analysis of binary images by computer.

As mentioned earlier, in binary images there are only two gray levels: black, the objects, and white, the background (or vice versa). The description and quantitative characterization of binary images involve the geometrical properties of objects and relationships among objects. This family of properties—not necessarily limited to binary images—is termed *texture*. Methods for describing textures have been developed for many different applications in pattern recognition, stereology, and mathematical morphology. However, many of the solutions proposed for either practical or theoretical problems have generated remarkable interdisciplinary overlaps.

After a brief look at these relationships, this chapter will discuss GIAPP's capability for parallel processing on a general-purpose minicomputer.

HISTORICAL BACKGROUND

The concept of applying television technology to quantitative image analysis closely followed the introduction of entertainment television. In the early 1950s Roberts and Young (1952) introduced the flying spot scanner, an electronic digitizer entirely under computer control that produces digital images from transparencies.

Following early attempts at designing character-recognition machines based on analog techniques, a first attempt to study pictures as sets of binary images produced by binary transformations was made after general-purpose computers became available, by Kirsch et al. (1957) and by Kirsch (1957), who developed some of the early ideas of Kovasznay and Joseph (1955) for processing two-dimensional patterns through scanning techniques.

Kirsch used a general-purpose computer (a SEAC with 1500 44-bit words of memory) for analyzing pictures stored in 176 × 176, or 30,976, binary digits (one picture occupied 704 words). He developed routines for computing the center of gravity of a pattern and translating the pattern rectilinearly so that the center of gravity is at the midpoint of the image, and computing a boundary or "first derivative" of an image in which each three-bit square is examined. In the latter routine if all nine bits are black the center bit is replaced by a white bit. This Kirsch termed to *custer*. He also computed "custer and complement" operations, elementary transformations that, as he remarked, produced results with far from trivial explanations.

Later, Uhr and Vossler (1961, 555) proposed a "program that generates, evaluates and adjusts its own operators." They showed examples of local 5 × 5 matrix operators used in character recognition. Many such operators, also termed *templates*, can be used in variously ordered fashion for achieving recognition. Uhr (1965) refers to a "template matcher," a program for a digital computer that matches the individual cells of the input matrix (the image) with the individual cells of a stored template. Not a "silly template," whose action is uncontrolled, such as randomly selected templates, will suffice, of course; neither will templates that are the patterns themselves (the objects to be recognized). Rather, what is needed are templates that are the "strokes" that compose the patterns (Rabinow, 1957). Uhr defines 1-tuple and n-tuple templates according to the number of cells supposed to match with the cells (pixels) in the image.

Moore (1968) initiated an organized approach to the analysis of binary images by computer in his description of a general-purpose computer program for image analysis, termed STRIP (Standard Taped Routines for Image Processing). He laid down the basis for the analysis of binary images as a relatively rapid computer method for the automated measurement of "structures."

After introducing the concept of *phase*, which represents the set of all particles of a single kind within an image, Moore wrote:

The phase concept is important since it allows the analysis of complex structures to be treated as a series of binary problems. In each treatment it is necessary only to consider the pattern formed by the particles of a single phase, against a black background occupied by all the other phases. Such a simplified treatment permits a computer equipped with an adequate group of Boolean operations to process an image at a rate around 200 times faster than is attainable when processing each image point as a numerical character. (Moore, 1968, 277)

He further added, "The basic operations generally referred to as Boolean operations or Boolean algebra, are implemented mainly by the very simple expedient of disabling the "carry" mechanisms which normally inform the machine that 6

14

+ 4 equals 10 rather than zero" (p. 294). Thus images can be considered as compact arrays of bits, and operations between images are computed with relatively high speed. Most general-purpose computers have this capability.

Moore described how sequences of logical operations performed on binary phase images, and "BIT OPerations" on them, can produce new images in which the number of pixels (or bits) that belong to the transformed objects are closely related to the morphometric measures sought. Through additional processing, the various objects are identified and separated in the images, and statistics are computed for characterizing the distribution of a large number of parameters. In support of his approach, Moore was probably the first to document experiments in which processes simulated in time by sequences of binary transformations provided criteria for predicting the behavior of the material analyzed: niobium-tin superconductor wires.

Three main lines of work have developed since those early approaches to binary image processing. One, directly following Moore's work, deals with the development of software such as FORTRAN programs, and of bit-manipulation routines written in machine language for general-purpose computers. A second line consists of the design of special hardware-built instruments connected to a television camera and monitor, which, in an interactive fashion, compute measures for selectively extracted binary patterns. The extraction is accomplished by means of semiautomatic gray-level slicing. A third line is the design of "parallel processors" and "pipeline processors," particularly fast computers that analyze gray-level images at real-time speeds. These complex and expensive computers process binary images as the simplest type of image data.

A recent software-based approach to image processing, inspired by Moore's concepts, is that of Rink (1976a, 1976b). Versatility and relative independence from capital costs for special equipment were the arguments in favor of the computationally slower approach.

Several specialized instruments, known as image-analyzing systems or image analyzers, have been built for the automatic detection of binary patterns from various types of image material (ore microscopy, metallography, cytology, etc.) and for precision morphological measurements of the detected patterns. In general, the image information is supplied by a wide range of input peripherals (optical microscopes, epidiascopes, slide or movie projectors, scanning electron microscopes, etc.) and passed onto the screen of a television camera. The electrical output from this camera passes into a closed-circuit television monitor and also to a detector unit, where signals from the camera emanating from the features that need to be measured are discriminated and selected from the rest of the signal. The output from the detector, consisting of pulses from the detected features, can be fed not only into the monitor (so that the operator can see which features he has detected) but also into a computer, which can be set to measure several geometrical properties (e.g., percentage area of a phase, number of detected features, their total projection and size distribution). Until recently the repertoire of measurements built as hardware in these instruments was severely limited.

15

A first model of one of the most reliable of these instruments, the Quantimet A, became available in 1963 (Fisher, 1967; Jesse, 1971), and a more defined modular system, the Quantimet 720, was introduced in 1970 (Coles, 1971; Fisher, 1971). Hougardy (1976) has published a summary review of image-analyzing instruments. Of particular interest was the model designed by Serra (1965, 1967, 1970) and described by Klein and Serra (1972). This model was constructed so that it could satisfy the statistical measurements of the theory of mathematical morphology for the study of geometrical aspects of two-dimensional sets (granulometric properties and structural properties).

From these experiences, a few years later, the Leitz Company in Germany built an image analyzer that was then termed TAS (Texture Analyzing System), see Muller (1974), Serra and Muller (1974), and Serra (1974).

Most image analyzers underwent continuous developments and expansions that required redesign of the modules for the specific tasks and additions of new options for added programmability. Perhaps the most recent developments are programmable systems similar to the AT4 and the TAS "second model" used by Chermant and Coster (1978) in France. Aside from having a rather extended programmability and being able to interface with large computers, these instruments have several memories for the storage of the detected images and their transformations. Although image analyzers are costly, they are well suited for routine measurements.

In a parallel processor either the entire image or a large part of it is processed at once in a number of elementary steps. The processing is organized in parallel (as opposed to sequential) fashion (most computers are sequential machines), thus making optimal use of the computer's capabilities in the minimum time. The concept of a two-dimensional parallel computer with a square module array was first introduced by Unger (1958, 1959), who referred to it as a computer oriented toward spatial problems for pattern recognition and detection. Unger proposed several simple programs that used local operations on binary images. The hexagonal array, instead, was proposed by McCormick (1963), who called it a "rhombic array," and hexagonal parallel pattern transformations were designed by Golay (1969) and discussed further by Preston (1971). PAX II, a first prototype of a parallel processing system, was described by Johnston (1970). At present *cellular logic* (or *neighborhood logic*) is the name given to a subdiscipline of computational geometry. It has ample applications in image processing, particularly in the design of parallel computers. In discussing state of the art, Preston et al. state:

Cellular logic refers to an operation performed digitally on an array of data $P(I,J)$ which is carried out so as to transform $P(I,J)$ into a new array $P'(I,J)$ wherein each element in the new array has a value determined only by the corresponding element in the original array along with the values of its nearest neighbors. The nearest neighbors configuration is called the "cell" and operations over arrays of identical cells are called "cellular logic." When one considers the implementation of cellular logic by an array of computers the logic and the memory associated with the cell define a processing element (PE). The array itself is then a cellular automaton (CA). (1979, 826)

Although programming was extremely difficult on the first parallel computers, it has become easier on the latest models, which are provided with high-level languages. Parallel computers, however, are expensive instruments dedicated to special purposes.

After the interest during the fifties and the sixties in the development of image processing methods for the analysis of binary images, the subject was practically abandoned during the seventies, except for applications related to the development of image analyzers.

These analyzers generated new needs for morphological measurements, which in turn favored the addition of general-purpose computers to the image analyzers. A renewed interest in the analysis of binary image information was generated partly as a result of the search for a theoretical background to probabilistic concepts and applications done by the French school of mathematical morphology. The work of stereologists in both the theoretical and practical aspects further supported the study of binary images.

The previous extract from Preston et al. (1979) summarizes well, in image-processing terminology, a logic that has its equivalent in the theory of transformations by means of "structuring elements," proposed in mathematical morphology, where the expression "hit-or-miss transformations" is frequently used. According to Serra (1976, 1978), the theory of mathematical morphology can provide the background for extrapolation and prediction when a stereological model appears necessary for the interpretation of geometrical properties of a structure. A recent book by Serra (1982) shows how set theory can be used to construct a picture algebra by developing criteria and models for image analysis. In particular, it considers the morphology of gray-tone functions—also studied by Goetcherian (1980)—and of random sets.

Most of the computations to be performed on images for morphological characterization are of a local nature, for example, edge detection, thresholding, thinning, skeletonizing, and counting. Thus a function is evaluated that, for each pixel, takes into account the values of a subset of neighboring pixels (not excluding the pixel itself). Local operators or neighborhood operators may be used for detecting both topological and geometrical features, in order to "understand" the image. A review of neighborhood operators in digital image processing has been made recently by Levialdi (1983).

ON SOME RELATIONSHIPS BETWEEN PATTERN RECOGNITION, STEREOLOGY, AND MATHEMATICAL MORPHOLOGY

Stereology is the study of the three-dimensional geometrical properties made from two-dimensional sections or projections through solid materials. It is interdisciplinary; covering fields as disparate as medicine, mineralogy, biology, metallurgy, mathematics, and statistics. Since about 1960, the main interest in

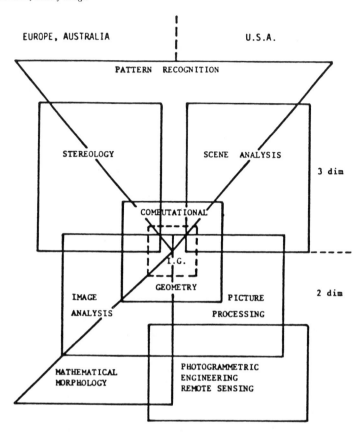

EUROPE, AUSTRALIA | U.S.A.

PATTERN RECOGNITION

STEREOLOGY   SCENE ANALYSIS   3 dim

COMPUTATIONAL

I.G.

GEOMETRY

IMAGE ANALYSIS   PICTURE PROCESSING   2 dim

MATHEMATICAL MORPHOLOGY

PHOTOGRAMMETRIC ENGINEERING REMOTE SENSING

Figure 3.1. Various branches in quantitative structural research. I. G. in the illustration refers to integral geometry. (*After G. Bernroider, 1978, The Foundation of Computational Geometry: Theory and Application of the Point-Lattice-Concept within Modern Structure Analysis, in Geometrical Probability and Biological Structure: Buffon's 200th Anniversary, R. L. Miles and J. Serra, eds., Springer-Verlag, New York, p. 155. Reprinted by permission.*)

pattern recognition has been in the development of automatic systems and machines. The instrumentation for data acquisition and preprocessing, some artificial intelligence methods, statistical approaches to image processing, and the development of picture languages, have been useful for the solution of stereological problems (Cheng, 1976; Chen, 1976). In addition, the different materials studied by stereologists have offered the opportunity and motivation for applying pattern recognition to a larger variety of everyday practical problems. Common to both stereology and pattern recognition, according to Pavel (1976), is the aim of identifying patterns from incomplete information. We can describe objects or images in terms of primitive (elementary) components and their composition, and also according to their local or global topologically invariant properties. A pattern is recognized through the detection of an equivalence (with regard

18

to a specific set of transformations), between the given objects and an element of a set of templates.

Bernroider (1978) points to probabilistic and computational geometry as a common methodological background between the various subdisciplines of quantitative structure analysis: stereology, image analysis and processing, pattern recognition, scene analysis, and mathematical morphology. He suggests that a connection might exist between picture-processing techniques, originally established in the United States over 20 years ago (Kovasznay and Joseph, 1955), and the "European school of Image Analysis and Stereology." This he considers important because the directions of structure analysis in the United States and Europe are different and seemed to be little aware of each other before 1977. Figure 3.1 portrays this situation.

LOGICAL OPERATIONS ON OR BETWEEN BINARY IMAGES

Boolean operations, Boolean algebra, or logical operations are computed between two computer words at the bit level. If we consider computer words as sets, each of n bits (8, 16, 32, 36, or 60, depending on the computer used) that can be either ON or OFF (0 or 1), we can obtain all the operations known in set theory by the symbols $\cup$ (union), $\cap$ (intersection), ^{c} (complement), and their combinations. In computer terminology, these operations are denoted by .OR., .AND., and .NOT., respectively.

Let us take two computer words and term them the sets A and B. We can write $A \cup B$, $A \cap B$, A^c, and B^c, which correspond to A.OR.B, A.AND.B, .NOT.A, and .NOT.B. Another operand that may be used is .EXOR. ("exclusive .OR."), symbolized by $\underline{\cup}$. This operation corresponds to the union of the non-overlapping subsets of two sets: $(A \cap B^c) \cup (B \cap A^c) = A \underline{\cup} B$, for simplicity in expression. When an element (pixel) a is contained in a set (image) A, it is denoted by ϵ A.

The operands .AND., .OR., .EXOR., and .NOT. are machine-dependent functions callable in most computer languages and can be used in rapid computations. In particular, machine-language routines can be written for logical operations between sets or arrays of words, for example, two rows of binary compressed images.

By performing these operations between successive portions (words or sets of words) of two images in corresponding subsequences of pixels, we can obtain logical operations between images at high speed. For example, if A and B consist of $1024 \times 1024 = 1,048,576$ bits, then on a 16-bit word computer these can be stored in $64 \times 1024 = 65,536$ words, and that many operations are needed to compute the picture resulting from a logical operation between A and B. In particular, if routines are used for the operations between arrays of 64 words, only 1024 calls to these routines will be needed for the operation between two images.

19

For illustration, Figure 3.2 shows several logical operations between (and on) two artificial binary images A and B, each of $10 \times 10 = 100$ pixels in size. The following ten operations are more commonly used: (1) A.AND.B, (2) A.OR.B, (3) A.EXOR.B, (4) .NOT.A, (5) .NOT.(A.AND.B), (6) .NOT.(A.OR.B), (7) .NOT.(A.EXOR.B), (8) A.AND.(.NOT.B), (9) A.OR.(.NOT.B), and (10) A.EXOR.(.NOT.B).

From the patterns in Figure 3.2 we can see how different relationships between two images can be expressed immediately by new images that result from logical operations. The results of three operations are identical, as can be seen in the figure, for $(A \underline{\cup} B)^c$, $A \underline{\cup} B^c$ and $B \underline{\cup} A^c$.

Logical operations can be combined with shifting (translation) operations to produce transformations that are used to characterize textures in binary images. Essentially, logical operations are parallel processes in which several bits (pixels) change their status at once in a single operation.

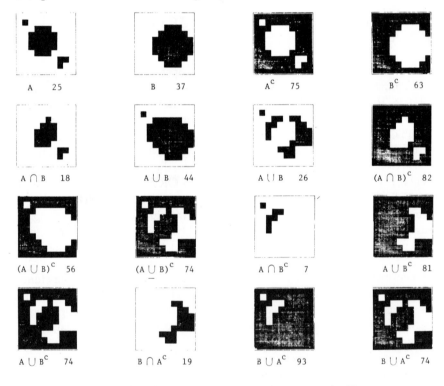

A 25	B 37	A^c 75	B^c 63
$A \cap B$ 18	$A \cup B$ 44	$A \underline{\cup} B$ 26	$(A \cap B)^c$ 82
$(A \cup B)^c$ 56	$(A \underline{\cup} B)^c$ 74	$A \cap B^c$ 7	$A \cup B^c$ 81
$A \cup B^c$ 74	$B \cap A^c$ 19	$B \cup A^c$ 93	$B \underline{\cup} A^c$ 74

Figure 3.2. Several logical operations between (and on) two artificial binary images, sets A and B, each of $10 \times 10 = 100$ pixels in size. The expressions below the images contain the following set symbols (and FORTRAN operators): $\cup$, union (.OR.); $\cap$, intersection (.AND.); ^{c}, complement (.NOT.); and $\underline{\cup}$, union of non-overlapping subsets (.EXOR.). Combinations of operations are also included, all of which are available in GIAPP as single operations. To the right of the logical expressions are the numbers of the black pixels in the binary images.

STRUCTURING ELEMENT TRANSFORMATIONS FOR THE ANALYSIS OF TEXTURES

In mathematical morphology, if we consider binary images as sets of points in the plane of the image, we can compute operations and transformations that measure their geometrical attributes. This can be done, for example, with the aid of appropriately designed elementary sets of points, which are used then as local operators.

A structuring element is a set of pixels that is swept across every pixel of an image whose ON-OFF status is changed according to the degree of matching in its corresponding neighborhood. Several structuring elements can be seen in Figure 3.3. When the central pixel of the structuring element happens to overlap a pixel in the image to be transformed, the pixel itself and the surrounding pixels identify its neighborhood for local computations.

Through the use of structuring elements, several geometrical properties of binary images can be measured and are immediately transformable into probabilistic statements. The set transformations in morphology may be generated by operations between sets developed by Minkowski (1911). The Minkowski sum of

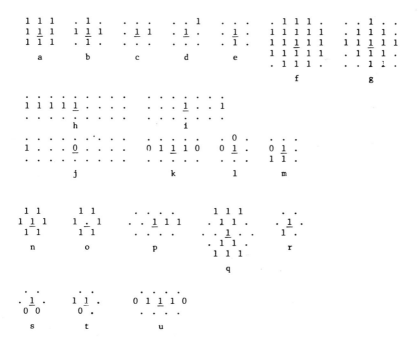

Figure 3.3. Some examples of structuring elements for a square raster, *a-m*, and for a hexagonal raster, *n-u*. The center pixels in the structuring elements are underlined. The dots identify the "don't care" positions; the 1's and the 0's identify the values of the neighbors used in the transformations.

21

sets A and B, denoted by $A \oplus B$, is all the pixels of the form $a + b$, in which "+" is the vector sum, a is contained in A, and b is contained in B ($a \in A$, and $b \in B$). The sum can also be denoted in terms of the logical expressions discussed earlier

$$A \oplus B = \bigcup_{b \in B} (\bigcup_{a \in A} (a + b))$$

Similarly, the Minkowski difference can be defined by

$$A \ominus B = \bigcap_{b \in B} (\bigcup_{a \in A} (a + b))$$

The sum is also referred to as *dilatation,* and the difference as *erosion* (Watson, 1975; Matheron, 1967, 1975) when the reflection of B respect to its center, $\check{B}$, is substituted to B. If the structure of B is symmetric respect to the center, then $\check{B}$ coincides with B. In Figure 3.3, symmetric structuring elements are: a, b, f, g, k, n, o, q, and u.

For example, in a square raster each element can be considered to be surrounded by eight neighbors. Suppose that our structuring element B is a square array of 3 × 3 black pixels (shown in Figure 3.3A). Each pixel in an image A is either black or white. By overlapping the structuring element with the pixels of the image, we can make the pixel corresponding with the centre of the structuring element black if at least one of its neighbors is black. Having termed this operation a dilatation, we can write for the transformed image $C = A \oplus B$.

An opposite transformation of A can be obtained by making the pixels in the image white if at least one of the pixels in the neighborhood is white. This operation is termed an erosion. The transformed image D is denoted by $D = A \ominus \check{B}$. These operations can be applied successively to the transformed images. The opening of an image by a structuring element is defined as an erosion followed by a dilatation; closing is a dilatation followed by an erosion. For the situations in which B is asymmetric with respect to its center, the opening is an erosion followed by a Minkowski sum, and closing is a dilatation followed by a Minkowski difference. Extensive application background to these concepts is explained by Serra (1975, 1976).

Figure 3.4 shows some transformations of a simple image A of 32 × 12 pixels by different structuring elements B1 to B4, in which the 1's and the 0's represent the valid points of the neighborhoods to be considered for the transformations; the dots indicate the "don't care" positions.

A transformation is evaluated according to the portion of pixels whose status has changed during the transformation. Logical operations between images before and after the transformations produce binary images of all the pixels that have changed status. In this manner, expressions of single geometric attributes can be extracted readily and displayed. The transformations described to the left of Figure 3.4 have been restricted to the following two situations:

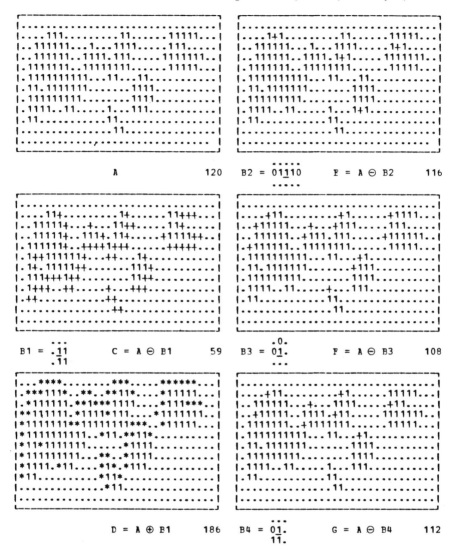

Figure 3.4. Erosions and dilatations of a 32 × 12-pixel artificial square raster binary image by different structuring elements B1 to B4. Structuring elements are shown below the transformations. Asterisks indicate white pixels in image A changed to black; crosses indicate black pixels changed to white. Dots indicate white pixels, and 1's black pixels, but in the structuring elements dots mean "don't care," and 0's mean white. The numbers of black pixels in each image or transformation appear below and to the right of the expression of transformation. In these structuring elements, both the 1's and the 0's identify the neighborhoods used in the transformations (both 1's and 0's should coincide, that is, match, with 1's and 0's in the image for computing the transformation). The center pixels in the structuring elements are underlined.

23

1. change status of pixel at the center to 1 if at least a 1 matches with the structuring element (dilatation), and

2. change status of pixel at center to 0 if at least a 1 mismatches with the structuring element (erosion).

The structuring elements used in Figure 3.3 were designed to be $(m) \times (n)$ arrays of binary pixels where (m) and (n) are odd integers ≥ 3. This was done to emphasize the similarity between structuring elements, binary neighborhoods, and templates or small images. In particular, the $(m) \times (n)$ structure is useful in programming the transformations with generalized structuring elements because the mapping of their center pixel through an image is considerably simplified.

Transformations can be generalized further to situations in which the matches have to occur for the 0's as well as for the 1's. The logic for the transformations then changes to:

3. if matching is perfect for all valid 0's and 1's in the structuring element, the image pixel at its center changes from 1 to 0 for an erosion, and

4. changes from 0 to 1 for a dilatation.

For such transformations, however, erosions can be obtained when the structuring element has a 1 at its center; for dilatations there must be a 0. The right side of Figure 3.4 shows erosions with several structuring elements according to transformation rule 3. In the terminology of mathematical morphology, such transformations are called *hit or miss transformations*. Here, for simplicity, it is preferred to maintain the two terms of erosion and dilatation for shrinking- or expanding-type of transformation, that is, for the situations in which the transformed images have less or more black pixels, respectively, than the original image.

The programming approach chosen has been generalized for both the square and the hexagonal rasters. It is illustrative to consider a few transformations of an artificial hexagonal image, which have been described by Serra (1978). Figure 3.5 shows how Serra's transformations are obtained in the algorithms developed in GIAPP.

Appendix C of this book provides some detail on "parallel" processing of the image in Figure 3.4. The examples shown in Figures 3.4 and 3.5 indicate transformations for which a structuring element is created as a small $(m) \times (n)$ binary image: it is the pattern within the structuring element (its configuration) that transforms binary images according to the degree of matching locally occurring. This concept can be further generalized for different degrees of matching and also for nonbinary images, as is generally done in the technique of template matching developed by Rosenfeld and Kak (1976).

REFERENCES

Bernroider, G., 1978, The Foundation of Computational Geometry: Theory and Application of the Point-Lattice-Concept within Modern Structure Analysis, in *Geometrical Probablity and Biological Structure: Buffon's 200th Anniversary*, R. L. Miles and J. Serra, eds., Springer-Verlag, New York, pp. 153-170.

24

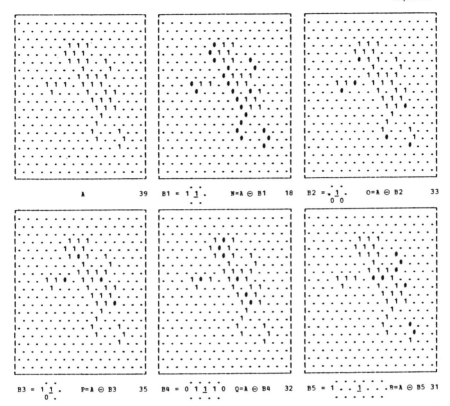

Figure 3.5. Erosion of an artificial 14 × 20 pixel hexagonal binary image by different structuring elements B1 to B5 (after Serra, 1978). Structuring elements are shown below the transformations. Symbols (#) indicate black pixels that changed into white pixels during the transformations. Dots indicate white pixels, and 1's black pixels. The numbers of black pixels in each image or transformation appear below and to the right of the expression of transformation. In these structuring elements, the 1's and the 0's indicate the part of the neighborhoods used in the transformations performed by B1 to B5; the dots indicate the "don't care" positions. The Center pixels in the structuring elements are underlined.

Chen, C. H., 1976, Theory and Applications of Imagery Pattern Recognition, in Proceedings of 4th International Congress for Stereology, Gaithersburg, Maryland, Sept. 4-9, 1975, *Nat. Bur. Stand. Spec. Pub. 431*, pp. 113-116.

Cheng, G. C., 1976, What Can Pattern Recognition Do for Stereology, in Proceedings of 4th International Congress for Stereology, Gaithersburg, Maryland, Sept. 4-9, 1975, *Nat. Bur. Stand. Spec. Pub. 431*, pp. 107-112.

Chermant, J. L., and M. Coster, 1978, Fractal Object in Image Analysis, in *International Symposium on Quantitative Metallography: Symposium Papers*, Nov. 21-23, 1978, Associazione Italiana di Metallurgia, eds., Florence, pp. 125-138.

Coles, M., 1971, Instrument Errors in Quantitative Image Analysis, *Microscope* **19:**87-103.

Fisher, C., 1967, An Image Analysing Computer, *Biomed. Eng.* **2:**351-357.

Fisher, C., 1971, The New Quantimet 720, *Microscope* **19:**1-20.

Goetcherian, V., 1980, From Binary to Grey Tone Image Processing Using Fuzzy Logic Concepts, *Pattern Recognition* **12**:7-15.

Golay, M. J. E., 1969, Hexagonal Parallel Pattern Transformations, *IEEE Trans. Comput.* **C-18**:733-740.

Hougardy, H. P., 1976, Automatic Image Analysing Instruments Today, in Proceedings of 4th International Congress for Stereology, Gaithersburg, Maryland, Sept. 4-9, 1975, *Nat. Bur. Stand. Spec. Pub. 431*, pp. 141-148.

Jesse, A., 1971, Quantitative Image Analysis in Microscopy — a Review, *Microscope* **19**:21-30.

Johnston, E. G., 1970, The PAX II Picture Processing System, in *Picture Processing and Psychopictorics*, Academic Press, New York, pp. 427-512.

Kirsch, R. A., 1957, Processing Pictorial Information with Digital Computers, *Am. Inst. Electrical Engineers*, vol. CP.-57-878.

Kirsch, R. A., L. Cahn, C. Ray, and G. H. Urban, 1957, Experiments in Processing Pictorial Information with a Digital Computer, *Proc. Eastern Joint Computer Conf.*, pp. 221-229.

Klein, J. C., and J. Serra, 1972, The Texture Analyser, *J. Microsc.* **95**:349-356.

Kovasznay, L. S. G., and H. M. Joseph, 1955, Image Processing, *IRE Proc.* **43**:560-670.

Levialdi, S., 1983, Neighborhood Operators: An Outlook, in *Pictorial Data Analysis*, R. M. Haralick, ed., Proceedings of 1982 NATO Advanced Study Institute, Bonas, France, August 1-12, 1982, Springer-Verlag, New York, v. F4, pp. 1-14.

McCormick, B. H., 1963, The Illinois Pattern Recognition Computer-ILLIAC III, *IEEE Trans. Elec. Comput.* **EC-12**:791-813.

Matheron, G., 1967, *Eléments pour une théorie des milieux poreux*, Masson et Cie., Paris, 166p.

Matheron, G., 1975, *Random Sets and Integral Geometry*, Wiley, New York 261p.

Minkowski, H., 1911, Theorie der Konvexen Korper insbesondere Begrundung ihres Oberflachen Begriffs, *Gesammelte Abh.* **2**:131-229.

Moore, A. G., 1968, Automatic Scanning and Computer Processes for the Quantitative Analysis of Micrographs and Equivalent Subjects, in *Pictorial Pattern Recognition*, G. C. Cheng, R. S. Ledley, D. K. Pollack, and A. Rosefeld, eds., Thompson, Washington, D.C., pp. 275-326.

Muller, W., 1974, The LEITZ-Texture-Analysing-System (LEITZ — T.A.S.), *Leitz Sci. Tech. Inf.* (suppl. 1) **4**:101-116.

Pavel, M., 1976, Projectors in Pattern Recognition Categories, in Proceedings of 4th International Congress for Stereology, Gaithersburg, Maryland, Sept. 4-9, 1975, *Nat. Bur. Stand. Spec. Pub. 431*, pp. 133-138.

Preston, K., Jr., 1971, Use of the Golay Logic Processor in Pattern-Recognition Studies Using Hexagonal Neighborhood Logic, in *Computers and Automata*, J. Fox, ed., Polytechnic Press, New York, pp. 609-623.

Preston, K., Jr., M. J. B. Duff, S. Levialdi, P. E. Norgren, and J-i. Toriwaki, 1979, Basics of Cellular Logic with Some Applications in Medical Image Processing, *IEEE Proc.* **67**:826-858.

Rabinow, J., 1957, Optical Coincidence Devices, U. S. Patent No. 2, 795, 705, June 11, 1957.

Rink, M., 1976a, A New, Fast and Storage-Saving Image Analysis Procedure for Investigating Individuals by a Digital Computer, in Proceedings of 4th International Congress for Stereology, Gaithersburg, Maryland, Sept. 4-9, 1975, *Nat. Bur. Stand. Spec. Pub. 431*, pp. 117-120.

Rink, M., 1976b, A Computerized Image Process for Isolating Individuals in an Originally Netted Patter, in Proceedings of 4th International Congress for Stereology, Gaithersburg, Maryland, Sept. 4-9, 1975, *Nat. Bur. Stand. Spec. Pub. 431*, pp. 155-158.

Roberts, F., and J. Z. Young, 1952, The Flying-Spot Microscope, *IEEE Proc.* **99**:747-757.

Rosenfeld, A., and A. C. Kak, 1976, *Digital Picture Processing*, Academic Press, New York, 457p.

Serra, J., 1965, Automatic scanning device for analyzing textures, J. Serra, inventor; Institute de Recherches de la Siderurgie Francaise (I.R.S.I.D.), assignee. First application in France, No. 1.449.059, July 2, 1965 (Patented in Belgium, Canada, England, Germany, Japan, Sweden, U.S.A.).

Serra, J., 1967, Buts et realization de l'analyseur de textures, *Rev. Ind. Miner.*, vol. 49, 9p.

Serra, J., 1970, Device for logical analysis of textures, J. Serra, inventor; Association pour la Recherce et le Developpement de Processus Industriels (A.R.M.I.N.E.S.) and J. Serra, assignees. First application in France, No. 70.21.322, June 10, 1970 (Patented in Austria, Canada, England, Germany, U.S.A. — application in Japan).

Serra, J., 1974, Theoretical Bases of the LEITZ-Texture-Analysing-System, *Leitz Sci. Tech. Inf.* (suppl. 1) **4:**125-136.

Serra, J., ed., 1975, *15 Fascicules de morphologie mathematique*, Centre de Morphologie Mathematique, Fontainebleau, France, 151p.

Serra, J., 1976, *Lectures on Image Analysis by Mathematical Morphology*, Cahier N-475, Centre de Morphologie Mathematique, Fountainebleau, July, 1976, 225p.

Serra, J., 1978, *One, Two, Three, Infinity*, in *Geometrical Probability and Biological Structure: Buffon's 200th Anniversary*, R. L. Miles and J. Serra, eds., Springer-Verlag, New York, pp. 137-152.

Serra, J., 1982, *Image Analysis and Mathematical Morphology*, Academic Press, New York, 610p.

Serra, J., and W. Muller, 1974, In Quantitative Image Analysis; the Problem of Resolution Errors Caused by Measuring Logic, *Leitz Sci. Tech. Info.* (suppl. 1) **4:**117-124.

Uhr, L., 1965, Pattern Recognition, in *Electronic Information Handling*, A. Kent and D. E. Taulbec, eds., Spartan Books, Inc., Washington, D.C., pp. 51-72.

Uhr, L., and C. Vossler, 1961, A Pattern Recognition Program that Generates, Evaluates, and Adjusts Its Operators, in *Proc. Western Joint Comput. Conf.*, pp. 555-569. Also in *Pattern Recognition, Theory, Experiment, Computer Simulations, and Dynamic Models of Form Perception and Discovery*, L. Uhr, ed., Wiley, New York, 1966, pp. 349-364.

Unger, S. H., 1958, A Computer Oriented Toward Spatial Problems, *IRE Proc.* **46:**1744-1750.

Unger, S. H., 1959, Pattern Detection and Recognition, *IRE Proc.* **47:**1737-1752.

Watson, G. C., 1975, Texture Analysis, *Geol. Soc. Am. Mem.* **142:**367-391.

CHAPTER 4

Example of Processing of Geological Data: Study of a Geological Map Pattern near Bathurst, New Brunswick

The experiments that will be treated in this chapter are of a descriptive nature: their aim is to introduce and explain the types of measurements and statistical estimates that are obtained from large regional geological maps. Concepts of spatial correlation of stratigraphic units quantified from geological maps will be illustrated by transformations of binary images digitized with a flying spot scanner. According to the theory of mathematical morphology, geometrical characteristics of binary images can be measured by structuring elements, and thus transformed into quantitative descriptions or geometrical probabilities associated with the patterns in the images.

The example of a map pattern in New Brunswick to be discussed was selected initially in 1977 for a study on the quantitative stratigraphic correlation of rock units and other features represented on geological maps (Agterberg and Fabbri, 1978). In 1979 this example was analyzed again by the author using some of the images previously studied but obtaining a number of new results, which also will be described here.

The approach aims to achieve a quantitative characterization and correlation of combinations of lithostratigraphic units coded from geological maps. Geological maps and ancillary information, also in map form, are constructed to aid in making interpretations, in planning new surveys or searching strategies, and in providing up-to-date syntheses of some geological processes for the prediction of other geological aspects.

This chapter has been adapted and expanded from F. P. Agterberg and A. G. Fabbri, 1978, Spatial Correlation of Stratigraphic Units Quantified from Geological Maps, *Comput. Geosci.* **4:**285-289.

29

Rock units on a map can be quantified with respect to a cell centered about an arbitrary point, and the average composition per cell can be determined by point counting, as described by Fabbri (1975). Mineral deposits tend to occur in specific types of regional environments. As exemplified by Agterberg et al. (1972), it is useful to attempt to define these environments so that they can be systematically compared to other environments in the same region.

Suppose that a random cell that contains one or more deposits of a specific type is termed a *deposit cell*. We can compare the statistical population of the deposit cell with the population of the random cell. If the random cell is sufficiently small, it tends to contain only one rock type, and only one deposit. A geological map, then, can be quantified by subdividing it into a regular grid of many small cells, and assigning to each cell the rock types that occur at its center. Suppose that rock type i has n_i cells in a region and that n_{id} of these cells contain a deposit. Then the probability that a random cell in rock type i is a deposit cell is equal to n_{id}/n_i. In addition, we have to consider that the probability of occurrence of a deposit also may depend on the spatial pattern of lithostratigraphic units at some distance — for example, 10 km — from the deposits.

Let us consider a geological map as a set of map units, each representing the shape and distribution, in at least two dimensions, of the lithostratigraphic units in the map. Over the entire surface of the geological map, each unit, distinguished by its color or graphical pattern, will occupy a given area. We can consider each map unit as a black pattern in a binary image, in which the white background represents the area occupied by all other map units together. Suppose that a small cell (a square, circle, hexagon, or octagon) is placed at random over the map area. The probability that such a random cell will "hit" the map pattern will be proportional to the area occupied by the pattern in the map; conversely, the probability that the random cell will "miss" the pattern will be proportional to the area occupied by all the other units together. If the cell is small enough, it will fall either entirely inside the pattern or entirely outside it. If such a cell is translated throughout the area to the intersections of a regular finely spaced grid, the proportion of the cell area occupied by the pattern and the probability that our cell will hit the pattern will be expressed by the same value. Let us consider the binary pattern for some ancillary data in map form, similar to the black-point (pixel) distribution pattern of mineral deposits. The probability that our random cell will contain one deposit is a function of the cell size and of the number and distribution of the points in the pattern. If the cell is so small that it can contain only one of the points at a time and is translated in a regular fashion as before, the areal proportion of the black-point pattern in the map and the probability of hitting the point pattern with our cell probe will have the same value.

When, in patterns systematically quantified from maps, we can identify conditions favorable to the occurrence of mineralizations (or other ancillary events), the geometric probabilities associated with these patterns can be combined with the probabilities associated with the distribution of mineral occurrences. This concept is the basis for developing predictive methods. A relatively simple application of this concept is presented here.

THE MAP PATTERN

The geology of the Bathurst area in New Brunswick has been described by Skinner (1974). Forty known massive sulfide deposits from this area are related genetically to the acidic volcanic rocks of the Tetagouche Group of the Middle to Late Ordovician. These acidic volcanics were coded from 2 miles to the inch geological maps (scale 1:126,720) from an experimental data base described by Fabbri, Divi, and Wong (1975). The pattern of acidic volcanics was drawn separately in black and white and photographed on 35-mm color film. The film was later digitized on the flying spot scanner as a set of 18,843 pixels on a square raster with a total of 324 × 320 = 103,680 binary pixels. The pixels are spaced 259 m apart in the north-south and east-west directions. The resulting binary image (shown in Fig. 4.1A) was obtained by thresholding of the gray-level image produced from the flying spot scanner. Such a transformation consisted of selecting a proper gray-level value or threshold, above which all pixels are considered black, while all remaining pixels are considered white. If the opposite convention is followed, a complementary binary image is obtained (shown in Fig. 4.1B). The square area digitized corresponds to 84 × 83 km. The same pattern was scanned also according to a hexagonal raster configuration as a set of 22,925 black pixels within an image of 313 × 349 = 109,237 pixels spaced 256 m apart. Each pixel in this second image can be considered as a hexagon with a minimum diameter of 256 m, with its center at that distance from the center of any adjacent neighboring pixel. (This image is shown in Figure 4.6A.)

According to Skinner (1974), the mapping of the Tetagouche Group is based on lithological units that have no stratigraphic significance because the group's stratigraphy and structure have not yet been determined. These units are characterized by sedimentary rocks, metabasalts, and rhyolitic rocks. The rhyolitic unit is shown in Figures 4.1A and 4.6A. The rhyolitic rocks are interpreted by the author as the youngest part of the group, surrounded by sedimentary rocks, the oldest part. The rhyolitic rocks, acidic volcanics, are possibly of ignimbritic (pyroclastic flow) origin.

EROSIONS AND DILATATIONS

In a square raster each pixel is surrounded by eight neighboring pixels. The eight pixels around any black pixel belonging to the image of Figure 4.1A, are either black or white. Suppose that if they are white, they are changed into black pixels: this would be termed *eight-neighbor expansion* (Rosenfeld and Kak, 1976), or *dilatation*. The result is a new image with 23,976 pixels, shown in Figure 4.1C. The difference between the two images, Figures 4.1A and 4.1C, consists of 23,976 -

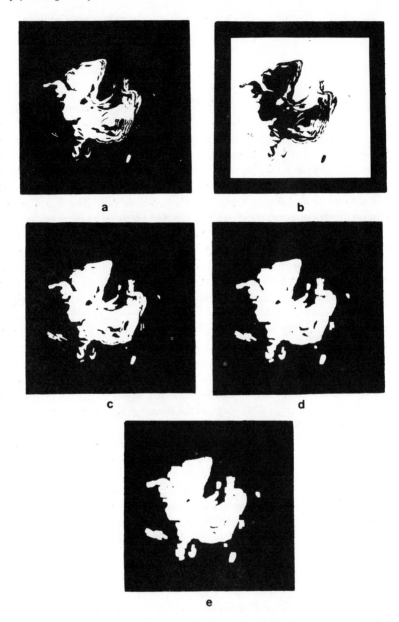

Figure 4.1. Erosions and dilatations of binary images of acidic volcanics in Bathurst area, New Brunswick. The original image A is shown in part A; there are 18,843 black pixels in this image. The complement of A (A^c) is shown in part B; there are 84,837 black pixels in this image. Writing B for the operator set of eight-neighbor expansion, the transformed images are as follows: (C) $A \oplus B$, 23,976 black pixels; (D) $A \oplus 2B$. 27,102

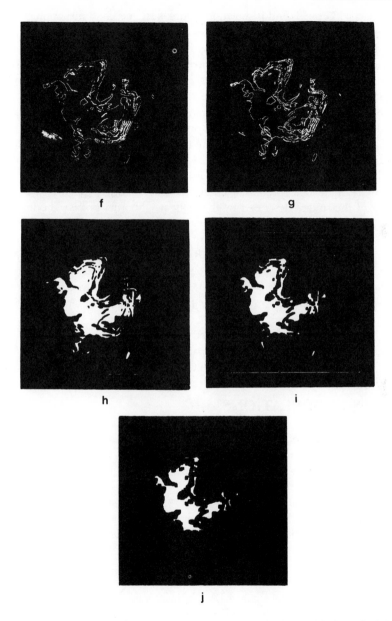

black pixels; (E) $A \oplus 3B$, 29,560 black pixels; (F) $(A \oplus B) \bigcap A^c$, 5133 black pixels; (G) $A \bigcap (A \ominus B)^c$, 5360 black pixels; (H) $A \ominus B$, 13,483 black pixels; (I) $A \ominus 2B$, 9990 black pixels; (J) $A \ominus 3B$, 7635 black pixels. Dimensions of the image are 84 km $\times$ 83 km; the north direction points upwards. (*After Agterberg and Fabbri, 1978.*)

33

18,843 = 5133 pixels and is shown in Figure 4.1F. A second dilatation gives the pattern of Figure 4.1D; a third produces the pattern of Figure 4.1E. The reverse operation, which consists of proceeding from the image in part A to that of part H, can be termed an *eight-neighbor shrinking*, or *erosion*. Three successive erosions of the original image in Figure 4.1A are shown in parts H, I, and J, respectively. The black pixels that turned to white in the first erosion are shown in Figure 4.1G.

To continue the discussion of these transformations, we can adopt the terminology developed by Serra (1976) and Watson (1975). Suppose that the original image in Figure 4.1A is a set A with measure (mes) A. This measure is the area and can be expressed either as mes $A = 18,843$ pixels or as mes $A = 1264.01$ km^2 (one pixel represents an area of 259×259 m). Let B be the operator set of the eight-neighbor-square logic (processing criterion). B has an origin that is located at the center of the square described by the eight neighboring pixels. The images of Figures 4.1C and 4.1D now can be represented as the Minkowski sums $(A \oplus B)$ and $(A \oplus 2B)$, respectively. In Agterberg and Fabbri (1978) a set nB was defined by induction with $nB = ((n - 1) B) \oplus B$ for $n = 2, 3, \ldots$. It is seen readily that operating on A with the set nB is identical to applying the successive operations $A \oplus nB = (A \oplus (n - 1) B) \oplus B$ for $n = 2, 3, \ldots$. Through Minkowski subtraction, we can write the patterns shown in Figures 4.1H, 4.1I, and 4.1J as $(A \ominus B)$, $(A \ominus 2B)$, and $(A \ominus 3B)$, respectively.

If ^{c} denotes the complement of a set with respect to the universal set T that consists of all the pixels in use, then the pattern in Figure 4.1B can be written A^c; the patterns of Figures 4.1F and 4.1G are $(A \oplus B) \cap A^c$ and $A \cap (A \ominus B)^c$, respectively.

CROSS-CORRELATIONS

A set C was formed by assigning each of the 40 volcanogenic massive sulfide deposits in the area to the pixels closest to it on the grid, with 259 m spacing used for the binary images of Figure 4.1A. C consists of 40 black pixels that can be subjected to successive dilatations by use of B. The images—sets $(C \oplus 4B)$, $(C \oplus 9B)$, and $(C \oplus 19B)$—are shown in Figures 4.2A, 4.2B, and 4.2C.

Because each pixel represents a cell that is 259 m on a side, the length of a cell generated by n dilatations is equal to $(2n + 1) \times 259$ m. Hence the cells obtained by 4, 9, and 19 dilatations of a single pixel are 2.33 km, 4.92 km, and 10.10 km, respectively. The latter two cell sizes can be used to approximate 5×5-km cells and 10×10-km cells, respectively. The binary images of Figures 4.2A, 4.2B and 4.2C can be intersected with that of Figure 4.1A. The resulting images—sets $(A \cap (C \oplus 4B))$, $(A \cap (C \oplus 9B))$, and $(A \cap (C \oplus 19B))$—are shown in Figures 4.2D, 4.2E, and 4.2F.

The erosion of Figure 4.1A can be continued from Figures 4.1H, 4.1I, and 4.1J onward until not a single black pixel remains. Likewise, the dilatations can be continued from Figures 4.1C, 4.1D, and 4.1E onward until most or all of the study area or image (set T) consists entirely of black pixels. The relative areas of the erosions and dilatations are shown in Figure 4.3. For dilatations each relative area can be interpreted as the probability $P(nB)$ with

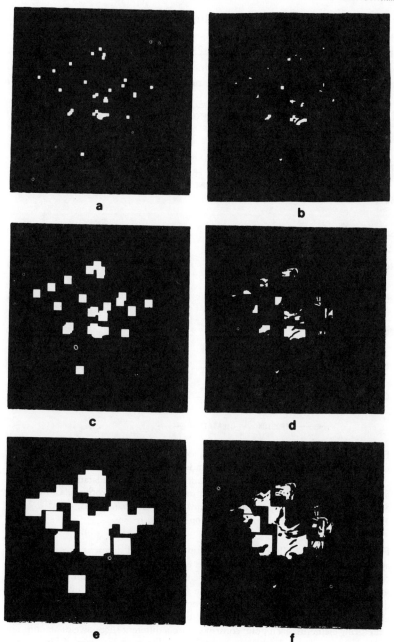

Figure 4.2. Dilatations of image set C for 40 deposit pixels and intersection of resulting sets with original image A in Figure 4.1A. *(A)* C ⊕ 4B, 2342 black pixels; *(B)* A ∩ (C ⊕ 4B), 1512 black pixels; *(C)* C ⊕ 9B, 8390 black pixels; *(D)* A ∩ (C ⊕ 9B), 4501 black pixels; *(E)* C ⊕ 19B, 26,654 black pixels; *(F)* A ∩ (C ⊕ 19B), 12,152 black pixels. *(After Agterberg and Fabbri, 1978.)*

35

$$P(nB) = \text{mes } A \oplus nB/\text{mes } T, \quad n = 1, 2, \ldots$$

that a random cell with side $((2n + 1) \times 259)$ m contains one or more black pixels belonging to the original pattern shown in Figure 4.1A. The probability $Q(nB)$ that a cell with size mes $(C + nB)$ contains no acidic volcanics is equal to $Q(nB) = 1 - P(nB)$.

Likewise, it is possible to measure the probability $P_d(nB)$ that a cell with side $((2n$

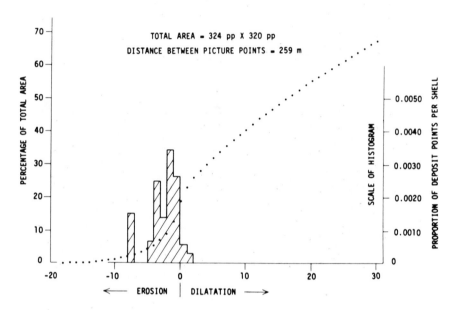

Figure 4.3. Percentage of total area occupied by the images obtained after successive erosions and dilatations. This histogram indicates the probability that a deposit pixel coincides with an arbitrary pixel in "shells added" after a single dilatation or "removed" after a single erosion. (*From Agterberg, F. P., and A. G. Fabbri, 1978 Spatial Correlation of Stratigraphic Units Quantified from Geological Maps, Comput. Geosci.* **4:**285-289, *Figs. 1-4, 8. Reprinted with permission by Pergamon Press Ltd.*)

Figure 4.4 (*at right.*) Probabilities that random cells, of different sizes and shapes approximating the circle, in the square raster and in the hexagonal raster configurations, contain one or more deposit pixels. Open circles indicate the areal proportions of the set T occupied by the binary image of the 40 deposit pixels after several successive dilatations by a square (*a*), hexagonal (*b*), octagonal (*c*) and (*d*) sets B, the structuring elements. The areal proportions are plotted against the horizontal diameters of the structuring elements; each point corresponds to a different successive dilatation. The vertical arrows point at the dilatations displayed in Figures 4.2 (square), 4.5 (octagons), and 4.6 (hexagons). Solid squares indicate areal proportions occupied by a hypothetical image of 40 black pixels such that the patterns of their dilatations do not overlap. Solid circles indicate the theoretical areal proportions of 40 non-overlapping circles of diameters equal to the horizontal diameters of the sets B used for the dilatations. The structuring elements and the elementary sets that produce the octagons are shown above the areal-proportion diagrams. Additional descriptions are in text.

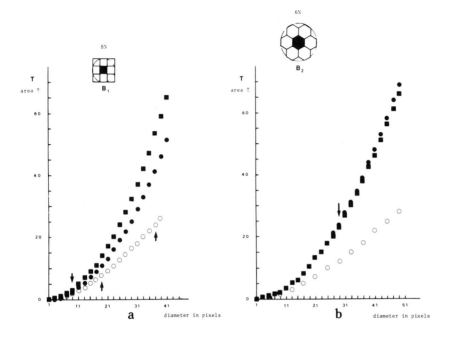

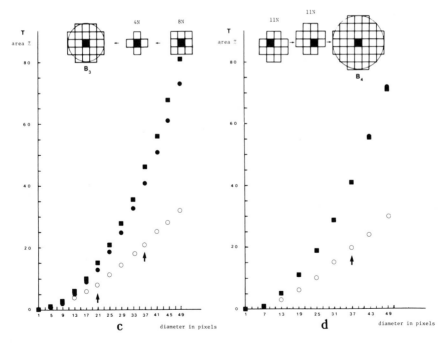

+ 1) × 259) m is a deposit cell containing one or more deposits, as it is shown in Figure 4.4A, because

$$P_d\,(nB) = \text{mes}\,(C \oplus nB)/\text{mes}\,T, \qquad n = 1, 2, \ldots.$$

Another practical result is as follows. A correlation between sets A and C can be carried out by determining how many deposit pixels are contained in the separate shells added to, or subtracted from, the original pattern of Figure 4.1A by means of dilatation or erosion. The original pattern itself contains 36 deposit pixels, or mes $(A \cap C) = 36$ pixels. The pattern of Figure 4.1G consists of 5360 black pixels and contains 14 deposit pixels. Hence the probability that an arbitrary pixel in this cell is a deposit pixel is equal to $14/5360 = 0.00261$. This probability is one of the proportions for separate cells shown in the histogram of Figure 4.3 The image of Figure 4.1I consists of 9990 pixels and has mes $((A \ominus 2B) \cap C) = 10$ pixels. This indicates that $36 - 10 = 26$ of the deposits (or 65 percent) occur in the zone identified as acidic volcanic rocks on the geological map, and within $(2 \times\sqrt{2} \times 259$ m $)= 733$ m of a contact between acidic volcanics and other rock types on this map. This zone may be relatively favorable for the occurrence of massive sulfide deposits. The probability that a random pixel in the zone is a deposit cell amounts to $26/(18{,}843 - 9990) = 0.00294$, which is about eight times greater than the probability (0.00039) that a random pixel in the entire study area is a deposit pixel. On the other hand, it is only about 1.5 times greater than the probability (0.00191) that an arbitrary pixel of the original binary image in Figure 4.1A is a deposit pixel.

As previously observed, 36 of the 40 deposit pixels (or 90 percent) coincide with the acidic volcanics, shown as black pixels in Figure 4.1A. A more general form of this ratio for a cell of size $((2n + 1) \times 259)$ m is:

$$M_{d1} = \text{mes}\,(A \cap (C \oplus nB)/\text{mes}\,(C + nB)$$

For the patterns shown in Figure 4.2, M_{d1} amounts to 0.646 (Figure 4.2D), 0.537 (Figure 4.2E), and 0.456 (Figure 4.2F). The ratio represents a weighted average proportion of acidic volcanics per cell for cells centered about the deposits. All these probabilities and ratios follow directly from Minkowski transformatioms of sets. Additionally, by computing logical operations between the sets, we can produce binary images that retain the spatial attributes that are hidden in the probability values.

TRANSFORMATIONS BY CIRCULAR ELEMENTS

In the Minkowski transformations discussed, the set B used was a 3×3 array of black pixels, representing a square (see Fig. 4.4A). Sometimes, particularly when computing measures related to distances in the two dimensions, the set B may be selected so that it approximates a circle as closely as possible (see Figs. 4.4*B*, *C*, and *D*). In a square raster a circle can be approximated by a pseudo-octagon. Two pseudo-octagons may be used for this purpose, one whose area exceeds that of the circle that has the same horizontal (or vertical) diameter, and a second whose area is less than that of the circle of the same horizontal diameter. The first type of octagon, shown in Figure 4.4*C*, is contained within a 5 × 5 array of pixels: its horizontal and vertical sides have lengths of 2 pixels, and its oblique sides have lengths of $\sqrt{2}$ pixels (1.4142). Its horizontal and vertical radius is 2.5 pixels. Transformations by this pseudo-octagon also can be obtained with two successive transformations by the two structuring elements of 4 and 8 neighbors (also shown in Fig. 4.4*C*). Figure 4.4*D* shows the second octagon, whose horizontal and vertical sides and radius are 3 and 3.5 pixels, respectively; its oblique sides are of length 2 × $\sqrt{2}$ (2.8284) pixels. Transformations by this pseudo-octagon can also be obtained with two successive transformations by two offset 11-neighbor sets (smaller octagons without a center pixel), also shown in Figure 4.4*D*.

For the hexagonal raster, such as the one used to produce the hexagonally scanned image of Figure 4.6*A*, a good approximation to the circle is obtained with the hexagon: this is shown in Figure 4.4*B*. Figure 4.4 shows as open circles the area proportions of the set *T* occupied by the binary patterns of the 40 deposit pixels after several successive dilatations by square, octagonal, and hexagonal sets *B* (structuring elements). Shown as solid squares are the areal proportions occupied by a pattern of 40 black pixels so that the patterns of their dilatations do not overlap. The functions of these proportions are compared with the theoretical area proportions, shown as solid circles, of 40 non-overlapping circles of diameter equal to the horizontal diameters of the four sets *B* used. As can be seen in Figure 4.4*A*, for increasing dimensions, the area of the square exceeds that of the circle more than the area of the octagon in Figure 4.4*C*. A closer approximation within the range of dilatations considered is obtained by the octagon in Figure 4.4*D* and the hexagon in Figure 4.4*B*.

The areal proportions in Figure 4.4 can be interpreted as probabilities that the different shapes, approximating circles of various diameters, translated at random throughout the image containing the pattern of 40 deposit pixels, hit one of these pixels. It also can be shown that the areas of hexagons and of octagons similar to the one in Figures 4.4*B* and 4.4*D* tend to depart from the areas of the corresponding circles with further increases in dimensions. The distribution pattern of the functions of the dilatations for our dot pattern, open circles in Figure 4.4, depends on the degree of clustering of the initial dot pattern, on the initial number of dots, and also on their location relative to the edge of the image set *T*.

The horizontal and vertical diameters of the octagons, in Figures 4.4*C* and 4.4*D*, generated for *n* dilatations, are equal to $((4n + 1) \times 259)$ m and to $((6n + 1) \times 259)$ m, respectively. The results of five and nine dilatations with the first octagon and of six dilatations with the second octagon correspond to diameters of 5439 m and 9583 m. The binary images in Figures 4.5*A*, 4.5*B* and 4.5*C*, corresponding to the above-mentioned octagonal dilatations of the deposit pattern, can be intersected with the geological pattern of Figure 4.1A. The resulting image sets (A ∩ (C ⊕ 5B3)), (A ∩ (C ⊕ 9B3)), and (A ∩ (C ⊕ 6B4)), where B3 and B4 are the octagons of Figures 4.4C and 4.4D, respectively, are shown in Figures 4.5D, 4.5E, and 4.5F. For these patterns, M_{d1} amounts to 0.530 (Fig. 4.5D), 0.468 (Fig. 4.5E), and 0.476 (Fig. 4.5F). The intersection of the sets in Figure 4.5B with the complement of the set in 4.5C is shown in Figure 4.5G to display the shape difference between the two types of pseudo-octagons that have the same horizontal diameter.

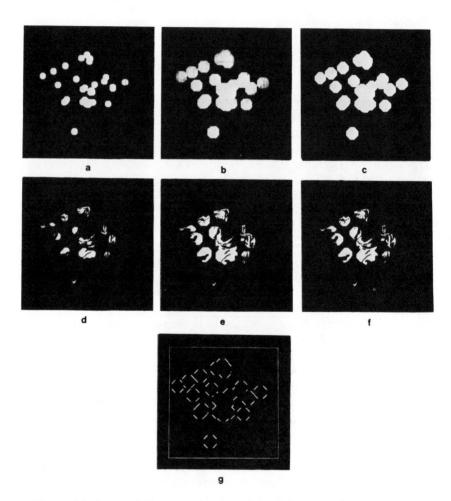

Figure 4.5. Octagonal dilatations of image set C for 40 deposit pixels, intersection of the resulting sets with the original image A in Figure 4.1A, and comparison of the octagons. (A) C ⊕ 5B3, 8736 pixels; (B) C ⊕ 9B3, 22,083 pixels; (C) C ⊕ 6B4, 20,164 black pixels; (D) A∩(C ⊕ 5B3), 4627 black pixels; (E) A ∩ (C ⊕ 9B3), 10,334 black pixels; (F) A∩(C ⊕ 6B4), 9593 black pixels; (G) (C ⊕ 9B3) ∩ (C ⊕ 6B4), 1919 black pixels in this image, which represents a comparison between dilatations of C by the two different pseudo-octagons when they have the same horizontal diameter.

40

HEXAGONAL TRANSFORMATIONS

Figure 4.6A shows the pattern of hexagonally scanned Middle-Upper Ordovician volcanic rocks. Fourteen dilatations of the hexagonal pattern of 40 deposits produced the image of Figure 4.6B: the set $(C \oplus 14B)$ in which the minimum diameter of the hexagon is equal to $((2n+1) \times 256)$ m, or 7424 m, for $n = 14$. The binary patterns of Figures 4.6A and 4.6B can be intersected, and the resulting image

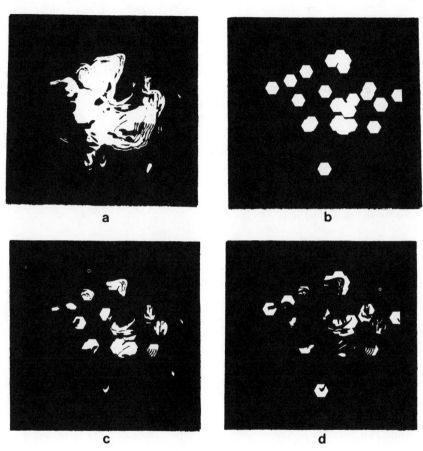

a

b

c

d

Figure 4.6. Cross-correlation between mineral deposits and the geological map pattern from a map that was scanned hexagonally on a flying spot scanner. *(A)* A binary image G of the map pattern of Middle-Upper Ordovician acidic volcanic rocks in Bathurst mining camp, New Brunswick. The image is $313 \times 349 = 109,237$ pixels, and is plotted hexagonally from hexagonal scanning. There are 22,925 black pixels in this image. *(B)* The result of 14 hexagonal dilatations D of the pattern of 40 volcanogenic massive sulfide deposits in the same area of G; there are 13,500 black pixels in this image. *(C)* The result of cross-correlating D and G: $D \cap G$. There are 7269 black pixels in this image. *(D)* The result of cross-correlating D and G^c: $D \cap G^c$. There are 6231 black pixels in this image. *(After Agterberg and Fabbri, 1978.)*

sets $(A \cap (C \oplus 14B))$ and $((C + 14B) \cap A^c)$ are displayed in Figures 4.6C and 4.6D, respectively. For the pattern in Figure 4.6C, M_{d1} amounts to 0.538; for the pattern in Figure 4.6D, where A^c is intersected, M_{d1} amounts to $1 - 0.538$, or 0.462.

Different shapes of structuring elements have been exemplified for the two geometrical arrangements of pixels: the square and the hexagonal raster. For isotropic transformations, the closest approximation to a circle is desirable at the given resolution. For anisotropic transformations, a rectangular structuring element or an approximation to an ellipse can be used. For direction-dependent transformations, linear structuring elements are used that point at given senses of direction. In some instances, the choice of structuring element is problem-dependent; for example, a square is preferable for relating geological data from maps with mineral deposit locations because it corresponds directly to the way the data are collected or digitized—according to the Universal Transverse Mercator (UTM) projection and coordinate system. A square raster representation of the pixels also is used for remotely sensed data.

The transformations described here have been computed using the generalized approach of GIAPP. Similar transformations also can be produced on special-purpose hardware-built instruments, such as the Quantimet 720 (octagons) or the Leitz TAS (hexagons). However, the degree of sophistication of those instruments does not yet allow a completely general approach to the technique of transformations by all types of structuring elements.

COVARIANCE MEASUREMENTS

Let us now consider briefly a few more types of transformations that can be used to characterize our spatial patterns further. For measurements in the plane at the crossings of a regular grid, a two-dimensional autocorrelation function can be computed that has an origin or a central point. Agterberg provided the theoretical background to the two-dimensional autocorrelation function as a direct extension of the one-dimensional function as follows:

Suppose that the elements of the $(m \times n)$ data matrix $\mathbf{X}$ are written as $X_{i,j}$ $(i = 0,1,\ldots,m - 1;$ $j = 0,1,\ldots, n - 1)$. The two-dimensional autocovariance function $C(r,s)$ is defined as

$$C(r, s) = \frac{1}{(m - r)(n - s)} \sum_{i = 0}^{m - r - 1} \sum_{j = 0}^{n - s - 1} (X_{i,j} - \overline{X})(X_{i + r, j + s} - \overline{X})$$

$$C(-r, s) = \frac{1}{(m - r)(n - s)} \sum_{i = r}^{m - 1} \sum_{j = 0}^{n - s - 1} (X_{i,j} - \overline{X})(X_{i - r, j + s} - \overline{X})$$

where $r = 0,1,\ldots M$, and $s = 0,1,\ldots, N$ [$\overline{X}$ is the mean value]. (Agterberg, 1974, 340, 341)

Each covariance $C(r,s)$ can be converted into an autocorrelation coefficient (r,s) by dividing by the variance $C(0,0)$. A model for the statistical covariance function of two-dimensional patterns (e.g., rock units on a map) which has been proposed by Agterberg (1977, 1981) was applied to the original pattern of acidic volcanics shown in Figure 4.1A. The geometrical covariance $K_a(l)$ satisfies $K_a(l) = $ mes $(A \ominus \breve{B})$, where A is the original pattern and $\breve{B}$ an operator set consisting of two points. One point is the origin of B and the other point occurs at a distance l in the direction$_2$. The symbol $\breve{\ }$ denotes reflection of a l set with respect to its origin. $K_2(l)$ is shown in Figure 4.7A for both the east-west and the north-south directions. Measurements are computed by translating a duplicate of the image being processed relative to it from a position of perfect overlap to a given number of positions in all directions, and computing the intersections or matching values for each position. The number of black pixels overlapping for the different shifts provides the geometrical covariance function. To obtain the corresponding statistical covariance, the values of Figure 4.7A were first increased by the factor mes T_0/mes $T_0 \ominus B$, where T_0 represents a square study area around A that measures 80 km on a side. The statistical covariances are obtained by subtracting m^2 from the corrected geometrical covariances where $m = $ mes A/mes T_0 is the proportion of the study area underlain by acidic volcanics. The statistical covariances (the values of Fig. 4.7B) were divided by the variance $C = m - m^2$, and this gave the autocorrelation coefficients r_l plotted along the vertical axis with the logarithmic scale in Figure 4.7C. The signal-plus-noise model, with $r_l = c$ $\exp(-p|l|)$ with $c = 0.80$ and $p = 0.106$ in the horizontal direction and $c = 0.76$ and $p = 0.080$ in the vertical direction, provides reasonably good fit. The two-dimensional covariance function shown in Figure 4.8 is isotropic by good approximation. The patterns corresponding to the values at east and south shifts ($\times$ 7 pixels) of $(0,0)$, $(5,0)$, $(0,5)$, and $(5,5)$, respectively, are shown in Figures 4.9A to 4.9D. The area of the light gray pattern divided by 10,000, is the value in the corresponding position of Figure 4.8.

HEXAGONAL CLOSINGS

A dilatation followed by an erosion is termed a *closing*. (For $B = \breve{B}$ a Minkowski subtraction produces the same transformation obtained by an erosion.) If produced by an isotropic set B, (a structuring element similar to the hexagon), this transformation can be used to describe distance relationships between the objects in a binary pattern. A hexagon B with a horizontal diameter of 5 pixels was used to produce first the set $(C \oplus B)$, dilatation of the pattern of 0 deposit pixels. This is shown in Figure 4.10A (see pg. 48). The 40 deposit pixels in the original pattern form 39 objects. This is so because two pixels occur side by side and therefore are considered a single object. The two pixels occur at a distance of 259 m. Figure 4.10A shows 24 objects after the hexagonal dilatation: this indicates that $39 - 24$, or 15, deposit pixels are less than $(4n + 1) \times 256$ m apart, which, for $n = 1$, is 1280 m. The pattern of Figure 4.10A was subjected to six closings (Fig. 4.10B to

43

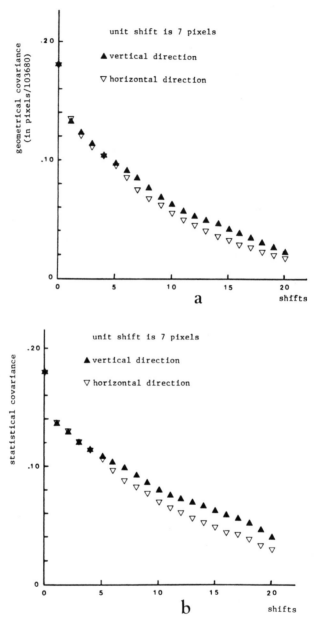

Figure 4.7. (*a*) geometrical and (*b*) statistical covariances for the east-west and north-south directions for the image of acidic volcanics in Figure 4.1*a*. The autocorrelation coefficient computed from the first ten geometrical covariances for the north-south direction (*c*) and the east-west direction (*d*) also are shown. The solid lines in (*c*) and (*d*) represent the theoretical autocorrelation function of the signal-plus-noise model. The subscript *l* refers to the distance between the two points in B as described in the text.

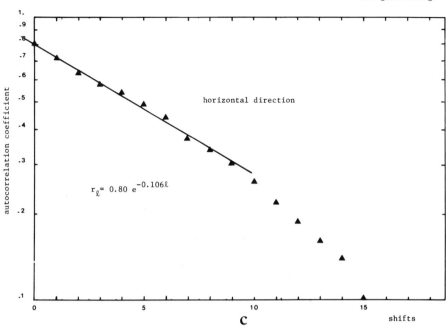

horizontal direction

$r_\ell = 0.80 \ e^{-0.106\ell}$

c   shifts

unit shift is 7 pixels

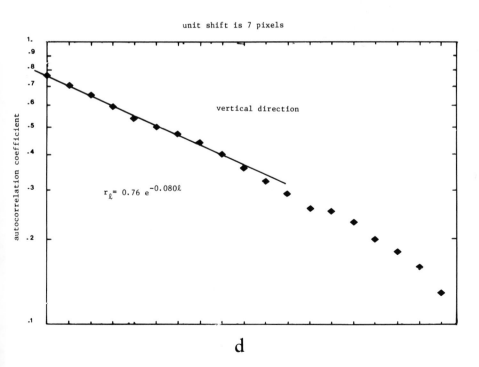

vertical direction

$r_\ell = 0.76 \ e^{-0.080\ell}$

d

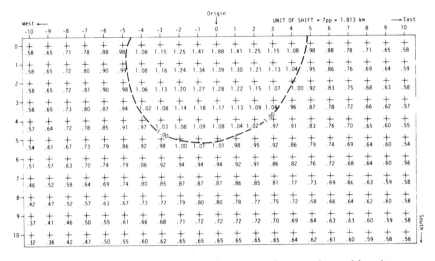

Figure 4.8. Two-dimensional geometrical covariance function obtained from binary image of Figure 4.1A (only lower half is shown). Values are in pixels/10,000. A contour for 10,000 pixels illustrates approximate isotropy. (*From Agterberg amd Fabbri, 1978. Reprinted with permission by Pergamon Press Ltd.*)

4.10G) consisting of one to six dilatations followed by the same number of erosions. This process of transformations creates "bridges" between objects whenever their edges are located at distances from other object edges that are less than or equal to $(2n + 1) \times 256$ m (where n is the number of closings). This happens if, during the nth dilatation, shapes are produced by merging object patterns that have their minimum diameter, in at least one of the three directions of the hexagonal raster, greater than the diameter of the n-dilatated hexagon used for the subsequent erosion. An object count after each closing iteration provides the distribution of the interobject or interparticle distances. In Figures 4.10A to 4.10G, the object numbers are 4, 22, 20, 20, 17, 17, and 17, respectively. The pattern of Figure 4.10h was produced by performing six dilatations followed by only five erosions of the pattern in Figure 4.10A. (One more erosion produced the pattern in Figure 4.10G.) This pattern shows the growing trend between clusters or groups of linked objects.

The problem of histogramming interparticle distances occurs in textural studies for ore minerals. In this essentially illustrative application, object counting is performed readily by eye. However, object counting in an hexagonal binary image can be performed automatically by the two structuring elements B2 and B3 shown in Figure 3.5. They provide the connectivity number defined as the number of objects minus the number of holes. This number can be used for constructing histograms of interparticle distances, as demonstrated by Serra and Verchery (1973).

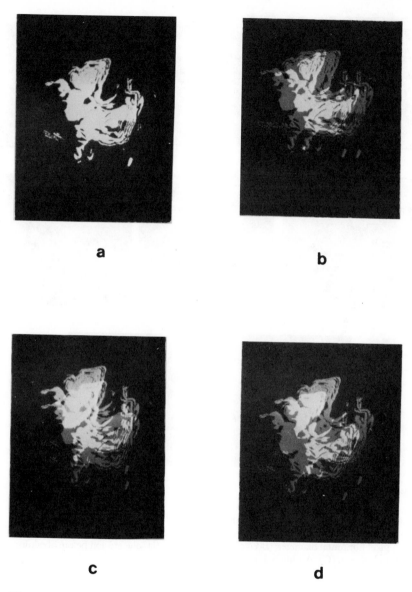

a

b

c

d

Figure 4.9. Shifting of a binary image during the computation of four of the geometrical covariance values shown in Figure 4.8. *(A)* (0,0) values for east and south shifts (X 7 pixels); *(B)* (5,0); *(C)* (0,5); and *(D)* (5,5). The area measured is the one of the light gray pattern.

47

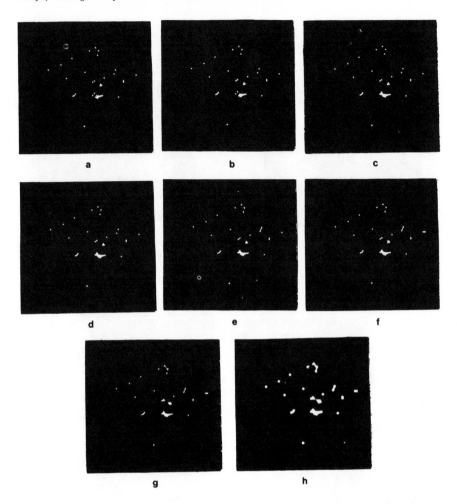

Figure 4.10. Hexagonal closings of a dilatated image of the 40 deposit pixels in the Bathurst area, New Brunswick, for describing the distance relationships between the hexagonal areas. (A) The 40 deposit pixels after a dilatation by a hexagon with a 5-pixel horizontal diameter; (B-G) one to six closings; (H) one erosion before obtaining the closing in (G). Additional information is in text.

REFERENCES

Agterberg, F. P., 1974, *Geomathematics*, Elsevier, Amsterdam, 596p.

Agterberg, F. P., 1977, Quantification and Statistical Analysis of Geological Variables for Mineral Resource Evaluation, Proc. Goguel Colloquium on Earth Sciences and Management, Orleans, France, in *Sciences de le Terre et Mesures, Memoir du B.R.G.M.*, no. 91, 1978, pp. 399-406.

Agterberg, F. P., 1981, Cell-Value Distribution Models in Spatial Pattern Analysis, in *Future Trends in Geomathematics*, R. G. Graig and Labovitz, eds., Pion, London, pp. 5-28.

Agterberg, F. P., and A. G. Fabbri, 1978, Spatial Correlation of Stratigraphic Units Quantified from Geological Maps, *Comput. Geosci.* **4:**285-294.

Agterberg, F. P., C. F. Chung, A. G. Fabbri, A. M. Kelly, and J. S. Springer, 1972, Geo-mathematical Evaluation of Copper and Zinc Potential of the Abitibi Area, Ontario and Quebec, *Geol. Surv. Can., Paper 71-41,* 55p.

Fabbri, A. G., 1975, Design and Structure of Geological Data Banks for Regional-mineral Potential Evaluation, *Can. Inst. Mining Metall. Bull.* **68:**91-98.

Fabbri, A. G., S. R. Divi, and A. J. Wong, 1975, A Data Base for Mineral Potential Estimation in the Appalachian Region of Canada, in *Report of Activities*, Part C, *Geol. Surv. Can., Paper 75-1C,* pp. 123-132.

Serra, J., 1976, *Lectures on Image Analysis by Mathematical Morphology*, Cahier N-475, Centre de Morphologie Mathematique, Fountainebleau, July 1976, 225p.

Serra, J., and G. Verchery, 1973, Mathematical Morphology Applied to Fibre Composite Materials, *Fibre Sci. Technol.* **6:**141-158.

Skinner, R., 1974, Geology of the Tetagouche Lakes, Bathurst, and Nepisiguit Falls Map-areas, New Brunswick, *Geol. Surv. Can. Mem. 371,* 133p.

Watson, G. C., 1975, Texture Analysis, *Geol. Soc. Am. Mem. 142,* pp. 367-391.

CHAPTER 5

Example of Processing of Geological Data: Analysis of a Portion of a Thin Section of a Metamorphic Rock

The descriptive experiments in this chapter introduce measurements and statistical estimates obtained from microscopic images of rocks in thin sections. Concepts of the quantitative characterization of microtextures (texture analysis) will be illustrated by transformations of binary images obtained from the boundaries of crystal-grain profiles digitized with a graphic tablet. The original drawing, shown in Figure 5.1*I*, of a thin section of a granulite from Otter Lake, Quebec, was taken from Kretz (1978). Fabbri and Kasvand (1978) and Fabbri (1980) describe part of the results which are treated here in more detail.

The following crystal phases occur in the thin section: calcic pyroxene, plagioclase, hornblende, and sphene. The grain-boundary image was processed in several steps, which included minor editing of the binary image produced from the vectors computed from the graphic-tablet output, line thinning, labeling of areas enclosed by contour lines, phase labeling, and extraction of binary images from the phase-labeled image. The purpose of the preprocessing was to obtain the patterns of the four different phases in the section as binary images in registration with one another. Similar preprocessing steps are described in Chapter 6.

The unit spacing between the pixels in the binary image is 0.0278 mm; the dimensions of the image are 180 pixels × 252 pixels. The boundaries of 398 individual crystals were digitized. Figure 5.1 shows the binary patterns obtained.

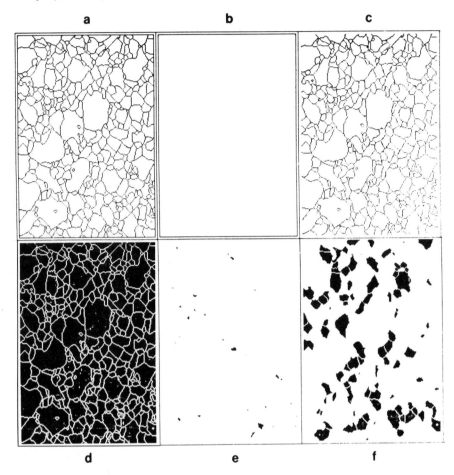

Figure 5.1. Binary images of a thin section of granulite from Otter Lake, Quebec. *(a)* Thinned boundary image (180 × 252 pixels) with rectangular frame added after thinning; *(b)* the rectangular frame at the edges of the image; *(c)* the boundary image without the frame; *(d)* image of all the grains together obtained by computing the complement of the image in *(A)*; *(e)-(h)*: binary images of sphene, calcic pyroxene, hornblende and plagioclase profiles, extracted from the image in *(a)*; *(i)* From *Kretz, 1978. Reprinted with permission of The University of Chicago and the author).*

EROSIONS, DILATATIONS, OPENINGS, AND CLOSINGS

Let us consider some representative transformations and measurements that are usually required with this type of rock material, and that illustrate the versatility of GIAPP.

During the interactive phase-labeling process, the pattern of all crystals adjacent to the frame is identified and extracted. This pattern then is used as a "mask,"

g   h   i

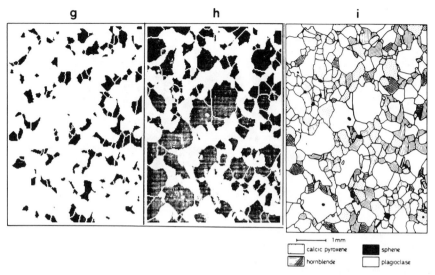

├———┤ 1 mm

☐ calcic pyroxene   ■ sphene

▨ hornblende   ☐ plagioclase

to eliminate those crystals from each crystal phase in which they occur. Figure 5.2A shows the binary pattern of all plagioclase crystals that were uncut by the edges of the picture or, in this instance, the frame. Several transformations of the binary pattern in Figure 5.2A were computed using the pseudo-octagonal structuring element of minimum diameter equal to five pixels, also shown in Figure 4.4A. The resulting patterns are shown in the remaining parts of Figure 5.2. Figures 5.2B and 5.2D show $(A \ominus B)$ and $(A \oplus B)$, single erosion and single dilatation, respectively, of the original set A (Figure 5.2A) by the set B, the octagon. An erosion provides a measure of the probability that an octagon with the shortest diameter of five pixels which is moved at random everywhere in the image space is contained fully within the plagioclase crystals. A dilatation provides a measure of the probability that the octagon hits, that is, falls partly or wholly on top of the plagioclase. These probabilities can be used, for example, to estimate the average circumference of the profiles of the plagioclase grains.

Figures 5.2C and 5.2E display an opening (an erosion followed by a dilatation) and a closing (a dilatation followed by an erosion). An opening can be used to estimate the grain-size distribution of the grains; a closing can provide information about interparticle distances. These measures should, however, take into account the irregularities of the boundaries of grains that have curvatures greater than can be approximated by the octagonal structuring element. Such aspects of the transformed images can be better displayed by logical operations between the original and the transformed images. Figures 5.2F to 5.2H show the results of the following operations: $A \cap (A \ominus B)^c$, $A \cap ((A \ominus B) \oplus B)^c$ and $((A \oplus B) \ominus B) \cap A^c$. A is the original set, and B is the octagonal structuring element set. Figure 5.2F shows black pixels that have become white in the octagonal erosion; Figure 5.2G shows black pixels that have turned from black to white during the octagonal opening of the image A in Figure 5.2A. The pixels added in the octagonal closing are displayed in Figure 5.2H instead.

53

a   b   c   d

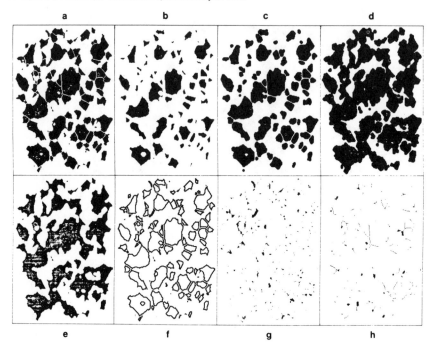

e   f   g   h

Figure 5.2. Minkowski-type operations with pseudo-octagonal structuring element on square raster image of plagioclase profiles from thin section of granulite. (*a*) Original image A of 180 × 252 pixels after elimination of all crystal profiles that were cut by the edge of the image; there are 17,344 black pixels in this image. (*b*) Octagonal erosion of A; there are 8996 black pixels in this image. (*c*) Octagonal opening of A; there are 15,935 black pixels in this image. (*d*) Octagonal dilatation of A; there are 25,939 black pixels in this image. (*e*) Octagonal closing of A; there are 18,355 black pixels in this image. (*f*) Result of logical operation $A \cap (A \ominus B)^c$; there are 8348 black pixels in this image. (*g*) Result of logical operation $A \cap ((A \ominus B) \oplus B)^c$; there are 1409 black pixels in this image. (*h*) Result of logical operation $((A \oplus B) \ominus B) \cap A^c$; there are 1011 black pixels in this image. (*From Fabbri, 1980. Reprinted with permission by Pergamon Press Ltd.*)

INTERPARTICLE DISTANCES AND PARTICLE COUNTING

As previously mentioned, closing transformations can be used to characterize interparticle distances. We have seen an application of closing for a point pattern in Figure 4.10, for a hexagonal image. For patterns in a square raster, several structuring elements can be used, such as the octagons of Figure 4.4 or the square. The larger is the structuring element, the broader are going to be the distance intervals used as unit. For example, horizontal and vertical distances of seven and five pixels and multiples of these can be measured using the octagons in Figures 4.4D and Figure 4.4C, respectively. For narrower intervals in a square raster, we can use the

54

square as the smallest isotropic shape approximation to a circle. For the square the shortest diameter or distance interval is three pixels.

The image of plagioclase crystals in Figure 5.1H was used to illustrate closing patterns and determinations of interparticle distances, by means of a square structuring element of increasing sizes (three and higher). In closing—that is, a dilatation followed by an erosion—the black pattern spreads. The growth transformation, however, cannot exceed the dimension of the original 180×252-pixel image. An erosion, on the contrary, is not affected by the image dimension, because it is a shrinking transformation. To avoid edge effects, the black pattern of plagioclases was mapped within a larger image of dimensions—200 pixels $\times$ 272 pixels—so that a width of ten pixels surrounds the edges of the original image. In the new larger image, the dilatated plagioclase crystals would reach the new image edge only at the tenth square dilatation. The patterns resulting from zero to five closing transformations are shown in Figures 5.3A to 5.3F, respectively. The initial number of crystals in the pattern of Figure 5.3A was 137; the number of connected objects in the five closed patterns is 22, 11, 5, 3, and 1, respectively. This is the information needed for histogramming the interparticle distances versus the number of crystals whose edges are less than a given minimum distance away from the closest crystal edge. The patterns of Figure 5.3 show how the bridges that develop connecting objects in a square raster image are elongated in either horizontal or vertical directions. In the hexagonal raster, unlike in the square raster, the bridges develop along the three main directions of the raster, as can be seen in Figure 4.10.

A set of transformations similar to the ones used for hexagonal images (B2 and B3 in Fig. 3.6) can be designed for square raster binary images, to compute the connectivity number for histogramming interparticle distances. The connectivity number (number of objects − number of holes) can be computed by erosions with the following two structuring elements:

$$B1 = \begin{matrix} . & 0 & . \\ 0 & \underline{1} & . \\ . & . & . \end{matrix} \quad \text{and} \quad B2 = \begin{matrix} . & . & . \\ 0 & \underline{1} & . \\ 1 & 1 & . \end{matrix}$$

where 1 indicate black pixels, 0 indicate white pixels, and dots indicate pixels not considered in the transformation.

To compute the connectivity number, it is necessary (a) to erode the original binary pattern A by B1, and compute how many black pixels have become white; (b) to erode A by the structuring element B2, and compute how many black pixels have become white; and (c) to compute the absolute difference between the counts in (a) and (b). The expression for this computation—the connectivity number, cn—can be written as follows:

$$| \text{mes } A - (\text{mes } A \ominus B1) - (\text{mes } A \ominus B2)| = cn$$

Figures 5.3G and 5.3H display the patterns of the black pixels that have become white after the erosions by the two elements mentioned above. They correspond to

55

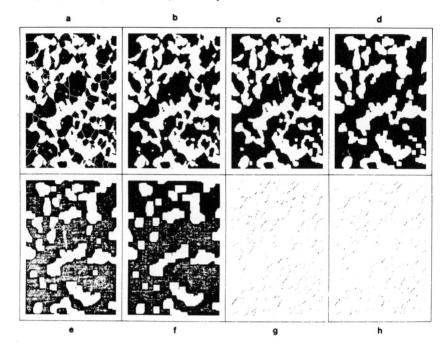

Figure 5.3. Characterization of interparticle distances, for the image of plagioclase profiles from the granulite, by successive closing operations. (*a*) The original image mapped into a larger image space of 200 pixels × 272 pixels; there are 137 profiles (objects) in this image. Parts *b* to *f* show results of one to five closing operations leaving 22, 11, 5, 3, and 1 "connected objects," respectively. (*g*) and (*h*) are the images of the black pixels that turned to white during the erosions of the image in (*a*) by the structuring elements

$$B1 = \begin{matrix} . & 0 & . \\ 0 & \underline{1} & . \\ . & . & . \end{matrix} \quad \text{and} \quad B2 = \begin{matrix} . & . & . \\ 0 & \underline{1} & . \\ 1 & \underline{1} & . \end{matrix}$$

respectively, for the computation of the "connectivity number."

the following expressions: $(A \cap (A \ominus B1)^c)$, and $(A \cap (A \ominus B2)^c)$. The set A, the original image, is shown in Figure 5.3A.

TRANSFORMATIONS BY LINEAR ELEMENTS

Linear structuring elements can be used to study the shape anisotropy of the grains in binary images. Figure 5.4A shows the binary image of calcic pyroxene for the same granulite. Figure 5.4B displays the black pixels that turned to white in the erosion by a horizontal linear element four pixels long extending to the right of the

56

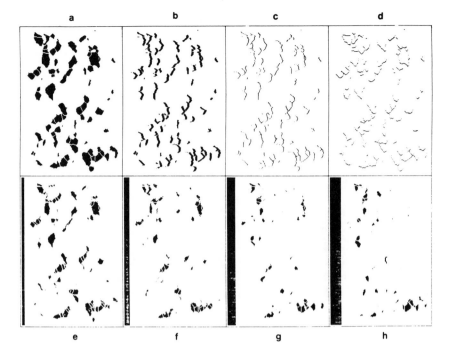

Figure 5.4. Minkowski-type operations with horizontal linear asymmetric structuring elements of different lengths on a binary image of calcic pyroxene from the same granulite as depicted of Figure 5.1, after elimination of the crystal profiles cut by the image edges. (a) Original image of calcic pyroxene crystal profiles; there are 6748 black pixels in this image. (b) Image of pixels that turned from black to white during the erosion of image A in (a) by the structuring element

$$
\begin{array}{ccccccccc}
\cdot & \cdot & \cdot & \cdot & \cdot & \cdot & \cdot & \cdot & \cdot \\
\cdot & \cdot & \cdot & \cdot & \underline{1} & 1 & 1 & 1 & 1 \\
\cdot & \cdot & \cdot & \cdot & \cdot & \cdot & \cdot & \cdot & \cdot
\end{array}
$$

There are 3516 black pixels in this image. (c) Image of pixels that turned from black to white during erosion of binary image A by the structuring element

$$
\begin{array}{ccc}
\cdot & \cdot & \cdot \\
\cdot & \underline{1} & 1 \\
\cdot & \cdot & \cdot
\end{array}
$$

There are 1030 black pixels in this image. (d) Image of the pixels which turned from black to white during erosion of image A by the structuring element

$$
\begin{array}{ccc}
\cdot & \cdot & \cdot \\
\cdot & \underline{1} & \cdot \\
\cdot & 1 & \cdot
\end{array}
$$

There are 860 black pixels in this image. (e)-(h) Computation of the linear horizontal geometrical covariance for right shifts of 5, 10, 15, and 20 pixels, respectively. The vertical black bands of widths of 5, 10, 15, and 20 pixels in (e) to (h) represent the loss in significance of the covariance values after shifting the entire images in the horizontal direction. The pattern in (e) to (h) simply represents the intersection between the image A and a shifted duplicate of the same image. *(Fig. 5.4a–d from Fabbri, 1980. Reprinted with permission by Pergamon Press Ltd.)*

center pixel. Figures 5.4C and 5.4D show the black pixels "eliminated" by the erosion of a linear right and downward element one pixel long. In stereology, such patterns are termed *horizontal* and *vertical intercepts*, respectively. These images and the measurements they provide describe quantitatively the average elongation of the grains of pyroxene in the vertical and horizontal directions. They can be considered the probabilities that horizontal and vertical trajectories placed at random throughout all the grains in the image space will intersect the right and the lower edges of the grains.

We also can represent the geometrical covariance as a transformation by a linear structuring element consisting of a set of two pixels, one black and one white, at a given distance from each other, and placed along a line oriented in a direction α. The origin of the set pair is at the 0 pixel. For direction $\alpha = 0$, we can consider the pair of pixels located in the horizontal direction. We can use this structuring element for producing a shifted pattern by erosion. If the origin 0 is at the pixel to the left, the pattern will be shifted leftward; if the origin is to the right, the pattern will be shifted toward the right. During the shifting operation a strip of our image pattern, elongated perpendicularly to the shift direction and of width equal to the length of the shift, will be lost from the edge of the image in the direction of the shift. A white band will replace it at the opposite edge of the image. The intersection between the shifted image and the original untranslated pattern is a new pattern from which we can compute the probability that a black pixel occurs at a distance equal to the shift in the direction of the shift. The pattern of Figures 5.4E to 5.4H, correspond to shifts of length 5, 10, 15, and 20 pixels, respectively. A black vertical band of width equal to the shift length has been added to identify the narrowing of the shifted images. The number of black pixels in each pattern has to be compared with the pixel count at the origin, that is, for a shift of zero pixels, or the proportion of black pixels in the original pattern over the entire universal image set $T0$ (of 180 pixels × 252 pixels). After each shift, however, new universal sets are produced, which are narrower than the set $T0$, such as the sets $T5$, $T10$, $T15$, and $T20$. Our probabilities will have to be scaled proportionally to those universal sets, smaller than $T0$ for the areas marked by vertical black bands in Figure 5.4.

COMPUTATION OF BOUNDARY LENGTH AND TRANSITION MATRIX

Let us now consider an application, on the patterns of Figure 5.1, to produce and display separate images of boundaries between different crystals and also of boundaries between crystals of the same type. As is sometimes done in textural studies in petrology, crystal-to-crystal transitions are coded for grain sequences along equally spaced traverses in order to compute transition matrices that can be compared to Markov chains for texture characterization; see, for example, Kretz (1969) and Whitten, Dacey, and Thompson (1975).

The approach consists of directly measuring the contact length for all contacts of the four minerals of the patterns in Figures 5.1E through 5.1H, and to compute the 4 × 4 transition matrix from these lengths. The images of Figures 5.1E through 5.1H were dilatated once by means of a square structuring element (3 × 3 pixels), and the resulting images were intersected with the image of boundaries of Figure 5.1C. This produced a separate boundary pattern for each crystal type. Figure 5.5A shows the boundary of hornblende.

Each boundary image then was intersected with any other one, which produced six partial boundary images, one for each possible pair of crystals. Figures 5.5B through 5.5D show the boundaries hornblende-sphene, hornblende−calcic pyroxene, and hornblende-plagioclase, respectively. The three partial boundaries so obtained for each crystal pair were cumulated, logical union operations with each other for every crystal type, and the resulting image was complemented and intersected with the boundary image of each crystal type. The results are images for the boundary patterns between crystals of the same type. Figure 5.5E shows all hornblende-hornblende boundaries.

From the different binary images of boundaries, the boundary lengths were computed as multiples of 1 for pairs of black pixels on the same row or column of the binary image, and as multiples of $\sqrt{2}$ for pairs of diagonal points. The 4 × 4 transition matrix computed is shown in Table 5.1, together with the contact length matrix, the percentage of contact length and area for each mineral, and the number of grains for each mineral in the image. This type of transition matrix describes the texture in the two dimensions by expressing the probability that each crystal is in contact with any other type of crystal and with crystals of the same type. The values in the table show a distinct tendency for unequal contacts over equal contacts. This pattern may provide evidence on genetic relationships between the minerals and on metamorphic recrystallization events of the rock, such as interfacial energy and nucleation.

It is of interest to consider briefly the problem of measuring the line lengths for the patterns of Figures 5.5A to 5.5E. Pairs of adjacent pixels with different orientations have been used as structuring elements for eroding the patterns in Figure 5.5D, the boundary between hornblende and plagioclase grains. The transformed patterns represent the proportions of boundary pixels adjacent to a boundary pixel to the right (Fig. 5.5F), below (Fig. 5.5G), to the lower right (Fig. 5.5H), and to the lower left (Fig. 5.5I). These four patterns, therefore, single out all point pairs at distances 1 from those at distances $\sqrt{2}$.

Furthermore, the patterns in Figures 5.5H to 5.5I correspond to the values of the geometrical covariance of the pattern in Figure 5.5D, for the shifts in the horizontal (−1, 0) and vertical (0, 1) directions. The auto-covariance arrays of values are listed in Table 5.2 for both the hornblende-plagioclase (Figure 5.5D) and the hornblende-hornblende patterns (Figure 5.5E). These arrays of autocorrelation values, however, underestimate the boundary lengths since no weight is given to the single points (which are eliminated by the erosions or by the unit shifts), and to the end points of line segments. This happens because the line patterns were measured out of their context. A better approximation of boundary lengths can be obtained by cross-

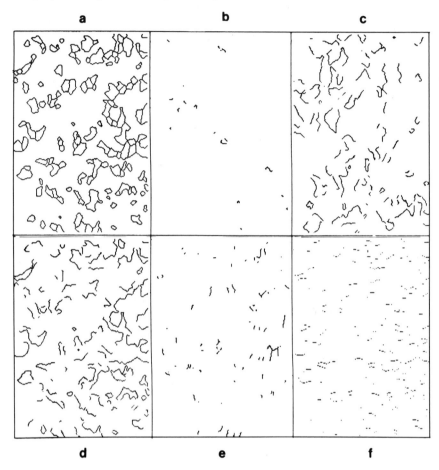

Figure 5.5. Computation of boundary length and transition matrix from the binary images of boundaries of profiles. (*a*) Binary image of hornblende profile boundaries; (*b*) binary image of sphene-hornblende boundaries; (*c*) binary image of pyroxene-hornblende boundaries; (*d*) binary image of plagioclase-hornblende boundaries; (*e*) binary image of hornblende-hornblende boundaries; (*f-i*) binary images of the pixels that turned from black to white during the erosions of the image in (*d*) by the following structuring elements:

$$
\begin{array}{cccc}
\begin{array}{ccc} \cdot & \cdot & \cdot \\ \cdot & \underline{1} & \cdot \\ \cdot & \cdot & \cdot \end{array}
&
\begin{array}{ccc} \cdot & \cdot & \cdot \\ \cdot & \underline{1} & \cdot \\ \cdot & 1 & \cdot \end{array}
&
\begin{array}{ccc} \cdot & \cdot & \cdot \\ \cdot & \underline{1} & \cdot \\ \cdot & \cdot & 1 \end{array}
&
\begin{array}{ccc} \cdot & \cdot & \cdot \\ \cdot & \underline{1} & \cdot \\ 1 & \cdot & \cdot \end{array}
\end{array}
$$

respectively, for detecting the type of black pixel adjacency in the four directions.

g   h   i

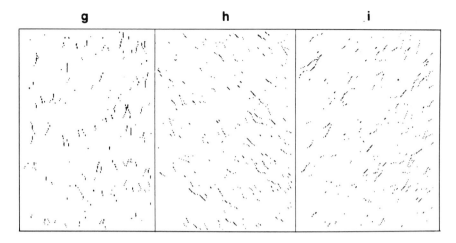

TABLE 5.1
Contact Length and Transition Matrices Computed for Boundaries of Sphene, Calcic Pyroxene, Hornblende, and Plagioclase

	Sphene	Pyroxene	Hornblende	Plagioclase	% contact length	% area	No. of crystals
Sphene	0.0000	0.0122	0.0109	0.0208	4.39	0.42	25
Pyroxene		0.0340	0.1496	0.2578	45.36	20.41	97
Hornblende			0.0474	0.3572	56.41	20.44	139
Plagioclase				0.1101	74.59	58.72	137

$$P = \begin{cases} 0.0000 & 0.2779 & 0.2843 & 0.4738 \\ 0.0269 & 0.0750 & 0.3298 & 0.5683 \\ 0.0193 & 0.2647 & 0.0839 & 0.6321 \\ 0.0279 & 0.3456 & 0.4789 & 0.1476 \end{cases}$$

correlating the patterns of Figures 5.5D and 5.5E with the pattern of Figure 5.5A, the total grain-boundary pattern for hornblende. The cross-correlation values are also listed in Table 5.2.

In these arrays, each value corresponds to distance units of 1/2 for horizontal and verical directions, and of $\sqrt{2}/2$ for oblique directions. Boundary lengths can be computed by multiplying the cross-correlation values by those factors and computing the sums of the eight values in the arrays. Such computation gives a length of 2786 units for the pattern in Figure 5.5D, and 442 units for that of Figure 5.5E. This approximation was used for the values of Table 5.1.

SKELETONIZATION BY LINE THINNING

Skeletonization is a transformation that detects all the pixels within objects that lie on, or are closest to, points at equal distance from opposite boundaries. Several

TABLE 5.2
Auto- and Cross-Correlation Values for the Computation of Boundary Lengths of Binary Images of Thinned Boundaries

Autocorrelation							Cross-correlation		
plagioclase/hornblende boundary			*hornblende/hornblende boundary*			*hornblende boundary* X *plagioclase/hornblende boundary*			
	-1	0	$+1$		-1	0	$+1$		-1	0	$+1$
$+1$	547	528	522	$+1$	41	163	67	$+1$	608	575	587
0	526	2195	526	0	16	353	16	0	551	2195	544
-1	522	528	547	-1	67	163	41	-1	565	565	599

Note: X indicates cross-correlated with.

algorithms have been proposed for such transformations, for either allowing object reconstruction or for facilitating object description and recognition. Algorithms exist for square raster images and for hexagonal raster images. The transformations are not necessarily limited to binary images, of course, but may be computed for gray-level images. Rosenfeld and Kak (1976) describe a few line-thinning algorithms. One type of skeletonization of binary images is the one produced by applying a line-thinning algorithm to patterns not necessarily made up of lines. Figure 5.6B is the set S, the "skeleton" pattern of calcic pyroxene. The set A is shown in Figure 5.6A after a closing transformation that joined all adjacent grains. The image was derived from the pattern in Figure 5.1F. In Figure 5.6C the pattern $(A \cap S^c)$ shows how closely the central pixels of the pyroxene crystal profiles are detected in the transformation. This type of transformation, however, is not computed in one single scan through the image; rather, it is iterative: processing continues and the image is transformed little by little, in cycles, until the process stops automatically because no further change is possible (analogous to the peeling of an onion). In the case of Figure 5.6A, the binary image was first expanded (one word per pixel), and then it was processed by a line-thinning algorithm adapted after Tamura (1975). The grains of the black pattern were thinned so that an 8-pixel connectivity between black pixels was maintained. The algorithm assigns labels of increasing negative values to the pixels belonging to consecutive shells (from the boundaries inward) to be transformed later into white pixels. The completely thinned pattern then was compressed, as shown in Figure 5.6B, where the apparent tendency of the grain cluster skeletons to be oriented preferentially in the vertical direction is clearly visible.

CONCLUDING REMARKS

As a general procedure, it is not particularly meaningful to apply many transformations to the material analyzed (the choice of transformations can be as wide as

TABLE 5.2 (*continued*)

			Cross-correlation				
hornblende boundary X hornblende/hornblende boundary			Distance Factors for Correlation Values			Lengths	
	−1	0	+1	−1	0	+1	
+1	64	176	96	+1 $\sqrt{2}/2$	$1/2$	$\sqrt{2}/2$	plag/horn = 2786
0	21	353	25	0 $1/2$	*	$1/2$	horn/horn = 442
−1	102	195	68	−1 $\sqrt{2}/2$	$1/2$	$\sqrt{2}/2$	

a   b   c

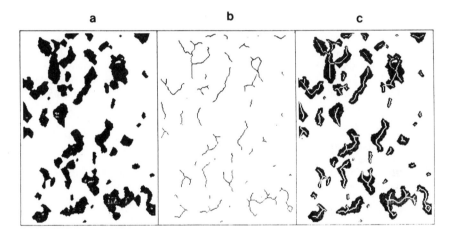

Figure 5.6. Skeletonization of a binary image by line thinning. (*a*) Image A of calcic pyroxene crystal profiles after a closing transformation by a 3 × 3 structuring element; (*b*) the image S of the skeletons of the pyroxene profiles (clusters) after line thinning; (*c*) result of the logical operation A ∩ S^c, which shows the performance of the line-thinning algorithm in the context of the original binary image.

one's imagination); rather, one should select or design those transformations that, depending on the problem at hand, are more likely to describe quantitatively the desired critical properties of the material analyzed.

Serra (1978) classified the transformations used in texture analysis according to the number of steps required: "One, Two, Three, ..., Infinity," that is, iterative transformations. Such approach can be extended logically from binary to gray-level images, as Goetcherian (1980) demonstrated. In addition, the neighborhood or structuring-element approach to transformations can be generalized further by modeling any given function within a neighborhood and assigning a given value to a certain pixel in it in relation to that function. Interestingly, the transformation approach, developed in this way, now has made a complete circle; thus so the old method of template matching has been rediscovered. Indeed a full exploitation of template-matching techniques can be done best by a specialized parallel or pipe-line processor. Although it may be remarked that so far there is nothing new under the sun in image processing, worthwhile new applications are being discovered every day that make it useful to continue to use the methodologies developed in the past.

REFERENCES

Fabbri, A. G., 1980, GIAPP: Geological Image Analysis Program Package for Estimating Geometrical Probabilities, *Comput. Geosci.* **6:**153-161.

Fabbri, A. G., and T. Kasvand, 1978, Picture Processing of Geological Images, in Current Approach to Stereological Problems, Proc. 5th Int. Congr. Stereol., Salzburg, Austria, Sept. 4-8, 1979, *Mikroskopie* **37:**431-436.

Goetcherian, V., 1980, From Binary to Grey Tone Image Processing Using Fuzzy Logic Concepts, *Pattern Recognition* **12:**7-15.

Kretz, R., 1969, On the Spatial Distribution of Crystals in Rocks, *Lithos* **2:**39-65.

Kretz, R., 1978, Distribution of Mg, $Fe^{**}+2$ and Mn in Some Calcic Pyroxene-Hornblende-Biotite-Garnet Gneisses and Amphibolites from the Grenville Province, *J. Geol.* **86:**599-619.

Rosenfeld, A., and A. C. Kak, 1976, *Digital Picture Processing,* Academic Press, New York, 457p.

Serra, J., 1978, One, Two, Three, Infinity in *Geometrical Probability and Biological Structure: Buffon's 200th Anniversary,* R. L. Miles, and J. Serra, eds., Springer-Verlag, New York, pp. 137-152.

Tamura, H., 1975, Further Considerations on Line Thinning Schemes, *Assoc. Electr. Comm., Japan, Prof. Note 66,* pp. 49-56 (in Japanese).

Whitten, E. H. T., M. F. Dacey, and K. Thompson, 1975, Markovian Grain Relationships of a Grenville Granulite, *Am. J. Sci.* **275:**1164-1182.

6

Techniques for Capturing Data from Geological and Ancillary Maps

PREVIOUS APPROACHES TO THE SYSTEMATIC ANALYSIS OF GEOLOGICAL MAP PATTERNS

Conventionally, geological maps are constructed so that they contain all the information that the geologist judges to be essential for description and interpretation, at the level of detail required by the scale of mapping. There is an increasing need to produce specialized maps that portray quantitative information for special purposes, such as mineral potential estimation, soil evaluation, geotechnical assessment, and environmental terrain classification. For those maps, different types of data have to be quantified, either manually, digitally, or electronically, and stored on files that can be processed by computer for subsequent analysis and visual display.

Several resources-related geomathematical projects have been developed by the Geological Survey of Canada since the mid-1960s. To produce estimates of the mineral potential for metal commodities in large areas of Canada, methods of statistical analysis of regional geological data, extracted from maps and retrieved from computer files on mineral deposits, were applied extensively by Agterberg et al. (1972). These applications required systematic compilation of special-purpose geological maps designed to reveal metallogenic features. The quantification and coding of regional geological information for resource estimates in the Appalachian region was described by Fabbri, Divi, and Wong (1975). For those projects, use was made of square grid patterns (equal-area cells) produced according to the Universal Transverse Mercator geographic projection (UTM). Transparencies of the grids were placed over the maps to be coded, and counts were made to express the areal proportions of all rock units occurring in each square cell of the grid.

As exemplified by Fabbri (1975), systematically quantified regional geological data can be useful for a variety of statistical applications. The data, however, which were coded manually and computerized for several thousands of 10 × 10-km cells, represented only part of the information contained in the maps. Only one cell size could be selected each time for manual compilation, and therefore the data units were either single cells or larger cells consisting of mosaics of individual unit cells. Such data have been used for correlating statistically the geological composition of the cells with the occurrence of mineral deposits (Leech, 1975). The main drawback of this procedure was the time and labor required in the manual collection and computer editing of the data. Additionally, in manually collected files, no direct information was coded on the two-dimensional shapes of the geological units or the lengths of their contacts. The spatial distribution of the different map units and the orientation of the lithological contacts were quantified only crudely, thereby severely limiting their subsequent analysis. This information can be of importance in deriving estimates of the probability of occurrence of mineralization, and in characterizing the spatial distribution of the mineral deposits in relation to the mapped variables.

A statistical approach to the analysis of geological variables from maps has been put forward by Switzer:

Complex spatial geological patterns may be regarded as realizations of random processes. The estimated parameters of such processes serve as convenient summary characterization of the observed geological patterns, and provide a basis for their classification and composition. The statistical properties of the estimates of process parameters, e.g., prevalence and patchiness, are related to the rate and methods of sampling, as well as to the model of the process itself. (1976, 124)

As Switzer proposed, the scale of the phenomena to which investigations of these relationships apply ranges from the texture of rocks in thin sections to satellite surface imagery. He paid particular attention to the study of models of spatial variation represented by simple two-color or multicolor patterns. Important applications in which the data are obtained characteristically at discrete locations are offered by soil samples, grab rock samples, points on a thin section, weather stations, digitized photographs and other imagery, and scoop samples from the ocean bottom. Discretely spaced sample data are used to estimate areal proportions (as in thin-section modal analysis), to determine pattern complexity, to make pattern reconstructions, and to obtain estimates of parameters of the pattern-generating process. Switzer proposes a method for estimating algebraically the spatial dependence of map patterns.

Some recent techniques of digitization of geological maps have been described by Bouille (1976), who used a graph-theory approach, and by Anuta, Hauska, and Levandowski (1976), who used a polygon technique to digitize geological maps for exploration purposes involving remote sensing and geophysical data. Srivastava

(1977) performed optical processing on structural contour maps in the form of "zebra maps," in which every other interval is black.

Not approached here are the problems associated with irregularly spaced data points and methods of interpolation of different types of data studied by Sampson (1975*a*, 1975*b*) and Robinson (1982).

DIGITIZATION, PREPROCESSING, AND PROCESSING OF LARGE REGIONAL GEOLOGICAL MAPS

Whenever the need arises to convert picture information into computer-readable form (that is, to digitize it), several alternative methods may be used. The nature of the data normally dictates the most appropriate method. Thus, with colored pictures, the data are usually raster-scanned using red, green, and blue filters. At each raster point the red, green, and blue color intensities are recorded as numbers. In this case the digitized data consist of three large matrices of numbers. In the case of gray-tone images, only one matrix is necessary because only the gray-level intensity is recorded. In the case of images containing lines only, such as the contour maps described in this chapter, an alternative to systematic raster scanning consists of some form of line tracing or line following. Given adequate equipment, automatic line tracing is feasible, but the lines may be traced manually with the aid of a specialized stylus and a graphic tablet digitizer. The tablet-and-stylus system is connected to a computer, which receives and stores the x-y coordinates of the stylus while the operator traces the lines on the map.

In raster scanning, the data consist of a matrix of numbers, whereas the data received from the tablet are a sequence of x-y coordinates, or "vectors," approximating the line being traced by the operator. The data structures stored in the computer for raster and vector mode are radically different. Whether the raster or the vectorial form of data is to be preferred is to a degree a matter of taste and available computer programs, but it is easier to solve some problems if the data are in raster form whereas other problems have easier solutions if the data are in vectorial form. Computer programs written for vectorial data cannot, however, use raster data, and vice versa. In principle, all data structures, vector or raster type, are equivalent, and if no loss of information occurs, they are convertible from one to the other, but the conversion may not be simple.

contour lines, etc.) are traced interactively and received by the computer in vector form. However, because GIAPP and the examples given here use raster-formatted data, vectorial data are converted to raster format for subsequent processing. In many situations raster format is desirable, but it may be wasteful in the required digital storage space, and does not allow scale or coordinate transformations to be carried out subsequently.

To prepare a map for digitizing, a systematic procedure must be followed (see Kasvand, Fabbri, and Nel, 1981). Because maps to be digitized may be larger than the available digitizing tablet, and to obtain suitable resolution (i.e., the maximum digital accuracy), rectangular subareas of each map, with no gaps and preferably no overlap between them, are selected and clearly marked. These rectangles are to fit within the active area of the graphic tablet. The final composite binary image will be rectangular. Each subarea or submap is digitized independently, and the optimum resolution obtainable from the tablet is preserved. A balance is needed between the number of submaps (fewer submaps means less work but less data) and the overall digital accuracy of the final product (more submaps means greater accuracy but more data). If many different large maps that cover the same area are to be digitized and processed, it is essential that the maps be in registration with one another, to facilitate crosscorrelations or matchings that imply point-to-point correspondence.

The following steps are involved in digitizing large geological maps:

1. A submap is mounted onto the graphic tablet with the sides preferably parallel to the coordinate system of the tablet, to minimize rotation. The tablet should have adequate grids to facilitate map mounting.

2. A relationship is then established between the rectangular submap and the tablet coordinates by pointing with the tablet stylus at the four corners of the submap and at the two ends of the scaling interval (see Fig. 6.1). This process can be repeated several times to obtain averages and minimize pointing errors.

3. The digitization is performed online: an operator traces the lines from the map, which the computer approximates with appropriate short line segments (vectors). During this tracing process—which is done on a graphic tablet by a pen-like cursor, or stylus—both the trace of the pen and the end points of the vectors computed are displayed on a graphic screen, as shown in Figure 6.2.

4. The raw digitized contour points, the corners of the rectangular subarea, the scaling interval in the map units, and the map identification code are stored on tape or disk for future usage.

5. A program divides the given scale interval into a desired number of resolution units (picture elements or pixels), specifies the ground distance (resolution) per pixel, and gives the overall size of the subpicture in pixels. The resulting constants or scaling values are used subsequently (see Fig. 6.3).

6. When all the rectangular submaps have been digitized and the raw data preserved, the data vectors are converted to raster form, in which the pixels underlying the digitized lines are represented as 1's, with 0's elsewhere. The raster data are thus stored in binary two-dimensional arrays or matrices. The procedure is equivalent to laying a transparent squared paper onto each submap and by marking each square as 1 if a contour line crosses the square, and as a zero 0, if no contour is in the square. The resulting marked overlay is a binary image of the map contours. The resolution of the image obviously depends on the size of the squares chosen. An example of vector-to-raster transformation is shown in Figure 6.4.

68

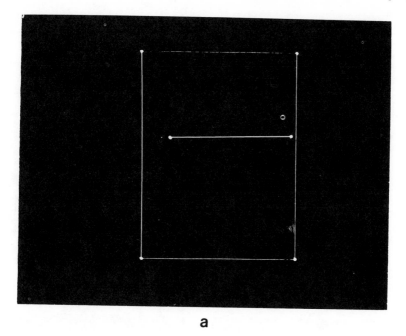

a

```
17
TABLT6        1      7      1      1      1
ENTER TOP LEFT CORNER OF PICTURE
ENTER TOP RIGHT CORNER OF PICTURE
ENTER BOTTOM RIGHT CORNER OF PICTURE
ENTER BOTTOM LEFT CORNER OF PICTURE
ENTER POINT #1 ON MAP SCALING INTERV
ENTER POINT #2 ON MAP SCALING INTERV
TYPE SCALING INTERVAL IN MAP UNITS:F
GIVE SINT:F10.0
50000.
     50000.
ISITOK? 1=YES, 2=NO, 3=ABORT:I1
1
RECTANGULARITY TEST ERRORS=  0.10142E+01 -0.99072E+00
RECTANGLE SLOPE ANGLE (DEGREES)=   359.1558
IG005:I2
```

b

Figure 6.1. (*a*) Plot on a Tektronix 611 video display of the four corners of the rectangle to be digitized and the scaling interval; (*b*) the conversation on the teletype related to this plot.

69

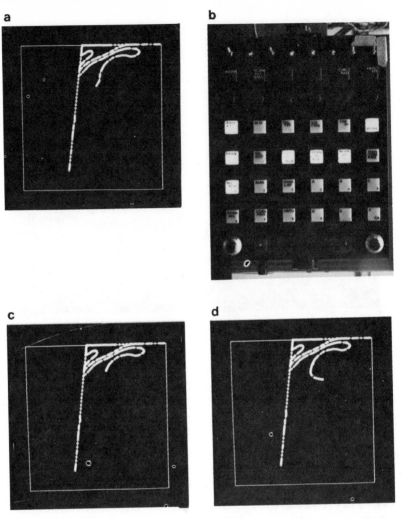

Figure 6.2. Vectors on a Tektronix 611 storage display unit during interaction for digitizing. (*a*) The digitized portion of a boundary: an error was made while digitizing. It was detected by looking at the display. Vector ends are marked by small circles, the position and length of each end point by a brighter dot. (*b*) The push-button box or command console. Editing online can be done by erasing, via push-button action, the last stored vector. (*c*) Display after erasing the erroneous vector and replotting the remaining stored vectors. The last seven vectors have been eliminated. (*d*) Display of the corrected boundary after removal of the wrong vectors and substitution by the corrected vectors.

```
18
TABLT7      300       1
NPIXEL,FPIXEL,DPIXEL =      300      167.2241          2.6757
AVERAGE DIMENSIONS:IDIMA,JDIMA =      380     502
CORNERS =  -511    668    505   -672
     -511.05408    668.97949    505.07739   -672.03271
T,CO,SI =     0.359663E+03    0.999983E+00   -0.587649E-02
16005:12
```

Figure 6.3. Interactive conversation that provides the constants for scaling values to be used in the processing. Three-hundred pixels is the number of resolution units desired, and the ground distance between pixels is computed from the scaling interval shown in Figure 6.1 as: 50,000 mi/(300-1) intervals = 167.2441 mi. Suggested dimensions for the sub-picture are 380 pixels horizontally and 502 pixels vertically. Tablet coordinates are also printed along with other relevant trigonometric values.

7. Normally the most complicated subpicture is selected first, and an adequate resolution for the binary image is determined. Steps 6 and 7 may have to be repeated on the chosen subpicture a few times. Once an adequate resolution is selected, it then is used for all the submaps of a set (see Fig. 6.5).

8. A composite binary image is created by inserting each binary subimage into its correct position. Due to its size, the large binary image (mosiac) may not be stored readily in the memory of the computer. The accuracy with which one submap can be aligned with another is on the order of one pixel (see Fig. 6.6).

9. Due to the resolution of the binary images, certain small areas may become closed, some adjacent contours may touch, and some breaks or other defects may occur. To ascertain that the pattern of the binary (large) image corresponds as well as possible to that of the original, the image should be printed and checked. Whenever feasible, the defects should be removed interactively by adding or deleting contour pixels via displays of enlarged portions of image details (see Fig. 6.7, 6.8, and 6.9). A hard copy of the area being edited can also be printed on the dot matrix printer.

10. The thickness of the contours, represented by the contour pixels in the binary image, is now reduced to a mimimum while peserving the connectivity of lines. This line-thinning process is a topology-preserving operation After this stage of preprocessing, the maps are represented by binary images containing thin (single-pixel) contours. The thinned form is more suitable for subsequent processing, because certain operations are simplified (see Fig. 6.9).

11. To identify each enclosed area in the image, a unique label or serial number now is assigned automatically to each area. The edges of the image also are considered as "contours," implying that all areas in the image are "enclosed." The image, however, is now no longer binary (see Fig. 6.10 and 6.11).

71

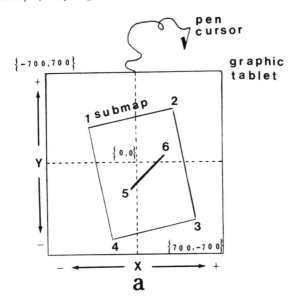

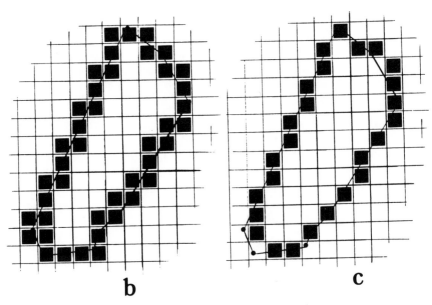

Figure 6.4. (*a*) Simplified diagram of the graphic tablet digitizer used; (*b*) fictitious example of vector-to-raster transformation, with white squares indicating white pixels, and black squares representing black pixels; and (*c*) transformation of black pixels representing a boundary in (*b*) into thin lines (line thinning).

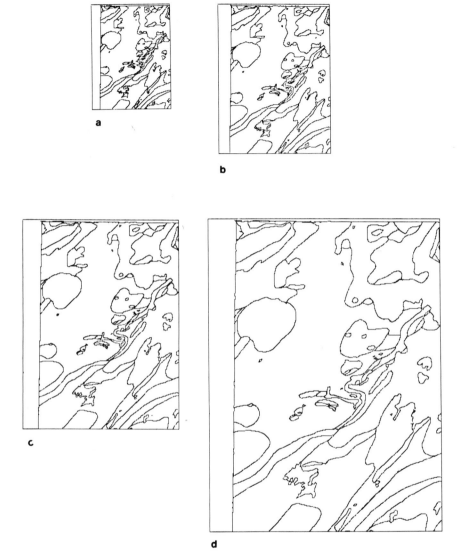

Figure 6.5. Binary images of a square portion of a geological map, obtained at different resolutions from the same vectors: *(a)* 1 pixel = 1000 m; *(b)* 1 pixel = 500 m; *(c)* 1 pixel = 375 m; *(d)* 1 pixel = 167 m, the resolution selected. These plots have been obtained on a Versatec dot matrix printer. The frames are not part of the picture data.

Figure 6.6. A mosaic of four independently digitized rectangular areas from a geological map, obtained on a Versatec dot matrix printer. The four images in (a) through (d) have dimensions of 380 pixels × 502 pixels each; the mosaic (e) has a dimension of 760 pixels × 1004 pixels.

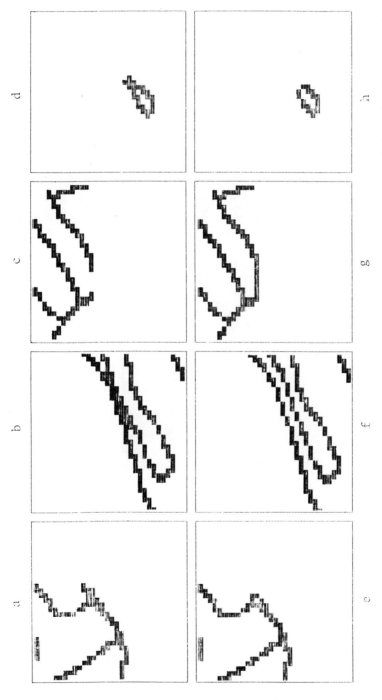

Figure 6.7. Occurrence and removal of digitizing defects from the binary images produced by transforming the vectors from the graphic tablet. Some examples of defects are as follows: (*a*) a small area that is closed; (*b*) two adjacent contours that touch; (*c*) a broken contour; (*d*) overshooting an enclosed contour. The removal of these defects is shown in (*e*) to (*h*). In these enlarged subpictures, each pixel is reproduced as an array of 6 dots × 6 dots.

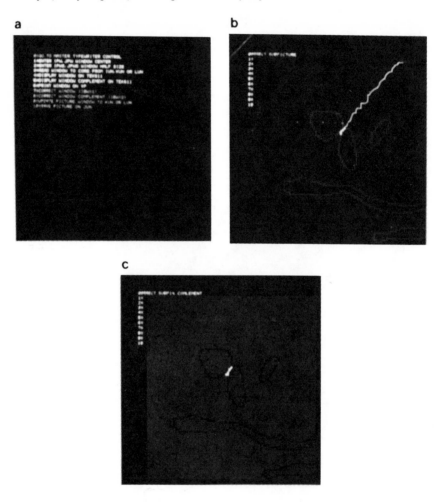

Figure 6.8. Interactive editing (addition or removal of black dots or pixels) of binary images via a special command console and online displays of enlarged portions of picture details. The command console is used to select the options tabulated on the Tektronix 611 screen in (*a*). A selected window is plotted in (*b*), and points are deleted by means of a pointer under the control of two wheels in the command console now used in the updating mode. The complement of the image window also can be displayed to "add" black pixels (i.e., to delete from the complement), as seen in (*c*). A hard copy of part of the area being edited is printed on a line printer (*d*). During these interactions the progressively edited versions of the binary image are stored on two disks, while the subarea being edited is in computer memory.

d

```
00000000000000000200000000000000000u0000000000000000000000000000800000000
333333333333333333333333333333333333333333333333333333333333333333333334
3334444444444555555555556666666666677777777778888888888999999999990
78901234567890123456789012345678901234567890123456789012345678901234567890
```

```
122...............................................................................
123...............................................................................
124...............................................................................
125...............................................................................
126...............................................................................
127...............................................................................
128...............................................................................
129............***************............................................
130..........*.............***...........................................
131.........**............**...........................................
132.........*.............*..............................................
133.........*.............*.................****......
134.........*.............*.................**...*.....
135.........**............*.................***...**....
136.........*.............*.................**....*....
137.........*.............*................**.....*...
138.........**.........*..***..............*.....**...
139..........*........*.*..**.............**....*....
140..........*........*..*...*...........**.....**..
141..........*.......*..*....*..........*......*..
142.........**.......*.*.....**..........*.....*..
143.........**.....***..*.....*.........*.....**..
144........**...***....*.....*..........**....*...
145.......****....**......**.........*....*...
146.............*.......*.........**..**.........
147.............*.......*.........***.........
148.............*......*..............................
149.............*......*..............................
150.............*......*..............................
151.............*.....**..............................
152.............*......*..............................
153.............**.....*..............................
154.............*......*..............................
155.............*......*..............................
156.............**.....*..............................
157............*......*..............................
158...........*.....**..............................
159..........**....*..............................
160.........**...*..............................
161..........*****..............................
162...............................................................................
```

```
00000000000000u00000000000000000m00000000000000000000000000000000000
333333333333333333333333333333333333333333333333333333333333334
3334444444444555555555556666666666677777777778888888888999999999990
78901234567890123456789012345678901234567890123456789012345678901234567890
```

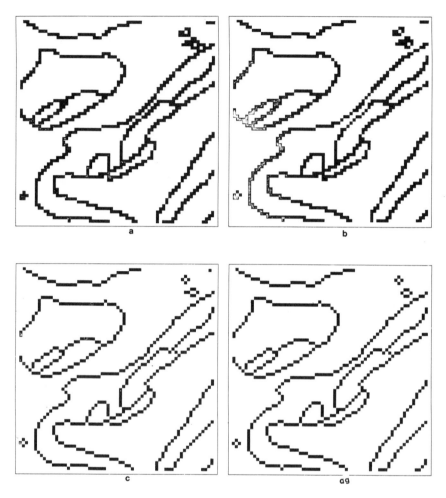

Figure 6.9. Editing and line thinning of the binary image of geological contours. A magnified portion is displayed before editing in *(a)*; after a first editing but before thinning, in *(b)*; after thinning in *(c)*; and after additional minor editing in *(d)*. A binary mosaic is displayed in *(e)* after a first editing but before thinning, in *(f)* after thinning, and in *(g)* after additional minor editing.

	502	503	504	505	506	507	508	509	510	511	512	513	514	515	516	517	518	519	520	521	522
383	0	0	0	0	0	0	0	0	0	0	0	1	0	0	0	0	0	0	1	0	0
384	0	0	0	0	0	0	0	0	0	1	1	0	0	0	0	0	0	1	0	0	0
385	0	0	0	0	0	0	0	0	1	0	0	0	0	0	0	0	0	1	0	0	0
386	0	0	0	0	0	0	0	0	1	0	0	0	0	0	0	1	1	0	0	0	0
387	0	0	0	0	0	0	0	1	0	0	0	0	0	1	1	0	0	1	0	0	0
388	0	0	0	0	0	0	1	0	0	0	0	0	1	0	0	0	0	1	0	0	0
389	0	0	0	0	0	1	0	0	0	0	1	1	0	0	0	0	0	1	0	0	0
390	0	0	0	0	1	0	0	0	0	1	0	0	0	0	0	0	0	1	0	0	0
391	0	0	0	0	1	0	0	0	1	0	0	0	0	0	0	0	0	1	0	0	1
392	0	0	0	1	0	1	0	1	0	0	0	0	0	0	0	0	0	1	0	0	1
393	0	0	1	0	0	0	1	0	0	0	0	0	0	0	0	0	1	0	0	0	1
394	0	0	1	0	0	1	0	0	0	0	0	0	0	0	0	1	0	1	1	0	0
395	0	1	0	0	1	0	0	0	0	0	0	0	0	0	1	0	0	0	0	1	0
396	0	0	1	1	0	0	0	1	1	1	1	0	1	1	0	0	0	0	0	0	1
397	0	1	0	0	0	0	1	0	0	0	0	1	0	0	0	0	0	0	0	1	0
398	1	0	0	0	1	1	0	0	0	0	0	1	0	1	1	1	0	0	0	1	0
399	1	0	0	1	0	0	0	0	0	0	0	0	1	0	0	0	1	0	1	0	0
400	1	0	0	1	0	0	0	0	0	0	1	1	0	0	0	0	0	1	0	0	0
401	1	0	1	0	0	0	0	0	0	1	0	0	1	0	0	0	1	0	0	0	0
402	0	1	0	0	0	0	0	1	1	0	0	1	0	0	0	0	1	0	0	0	0
403	0	1	0	0	0	0	1	0	0	0	0	0	1	0	0	1	0	0	0	0	0
404	0	1	0	0	0	1	0	0	0	0	0	1	0	1	1	0	0	0	0	0	0
405	0	1	0	1	1	0	0	0	0	0	1	0	0	0	1	0	0	0	0	0	0
406	0	0	1	0	0	0	0	0	0	1	0	0	0	0	0	1	0	0	0	0	0
407	0	0	1	0	0	0	0	0	0	1	0	0	0	0	0	1	0	0	0	0	0
408	0	1	0	0	0	0	0	0	0	1	0	0	0	0	0	1	0	0	0	0	0
409	0	1	0	0	0	0	0	0	1	0	0	0	0	0	0	1	0	0	0	0	0
410	0	1	0	0	0	0	0	0	1	0	0	0	0	0	0	1	0	0	0	0	0
411	0	1	0	0	0	0	0	0	1	0	0	0	0	0	0	1	0	0	0	0	0
412	0	1	0	0	0	0	0	0	1	0	0	0	0	0	0	0	1	0	0	0	0
413	0	0	1	0	0	0	0	0	1	0	0	0	0	0	0	0	1	0	0	0	0
414	0	0	1	0	0	0	0	0	1	0	0	0	0	0	0	0	1	0	0	0	0
415	0	0	1	0	0	0	0	0	1	0	0	0	0	0	0	0	0	1	0	0	0
416	0	0	1	0	0	0	0	0	1	0	0	0	0	0	0	0	0	0	1	0	0
417	0	0	1	0	0	0	0	0	1	0	0	0	0	0	0	0	0	0	0	1	1
418	0	0	1	0	0	0	0	0	1	0	0	0	0	0	0	0	0	0	0	0	0
419	0	0	0	1	0	0	0	0	1	0	0	0	0	0	0	0	0	0	0	0	0
420	0	0	0	1	0	0	0	0	1	0	0	0	0	0	0	0	0	0	0	0	0
421	0	0	0	1	0	0	0	0	1	0	0	0	0	0	0	0	0	0	0	0	0
422	0	0	0	1	0	0	0	0	1	0	0	0	0	0	0	0	0	0	0	0	0
423	0	0	0	1	0	0	0	0	1	0	0	0	0	0	0	0	0	0	0	0	0
424	0	0	0	0	1	0	0	0	1	0	0	0	0	0	0	0	0	0	0	0	0
425	0	0	0	0	1	0	0	0	1	0	0	0	0	0	0	0	0	0	0	0	0
426	0	0	0	0	0	1	0	0	1	0	0	0	0	0	0	0	0	0	0	0	0
427	0	0	0	0	0	0	1	1	0	0	0	0	0	0	0	0	0	0	0	0	0
	502	503	504	505	506	507	508	509	510	511	512	513	514	515	516	517	518	519	520	521	522

Figure 6.10. A binary expanded 21 × 45-pixel portion of the image of thinned geolocigal contours. Boundary pixels are represented by 1's, and the areas contoured by 0's.

	502	503	504	505	506	507	508	509	510	511	512	513	514	515	516	517	518	519	520	521	522
383	276	276	276	276	276	276	276	276	276	276	276	1	389	389	389	389	389	389	1	401	401
384	276	276	276	276	276	276	276	276	276	1	1	389	389	389	389	389	389	1	401	401	401
385	276	276	276	276	276	276	276	276	1	389	389	389	389	389	389	389	389	1	401	401	401
386	276	276	276	276	276	276	276	276	1	389	389	389	389	389	389	389	1	1	401	401	401
387	276	276	276	276	276	276	276	1	389	389	389	389	389	1	1	401	401	401	401	401	401
388	276	276	276	276	276	276	276	1	389	389	389	389	389	1	420	420	420	420	1	401	401
389	276	276	276	276	276	1	389	389	389	389	1	420	420	420	420	420	1	401	401	401	401
390	276	276	276	276	1	389	389	389	389	1	420	420	420	420	420	420	1	401	401	401	401
391	276	276	276	276	1	389	389	389	1	420	420	420	420	420	420	420	1	401	401	401	401
392	276	276	276	1	430	430	1	420	420	420	420	420	420	420	420	420	1	401	401	1	
393	276	276	1	430	430	430	1	420	420	420	420	420	420	420	420	420	1	401	401	1	
394	276	276	1	430	430	1	420	420	420	420	420	420	420	420	420	1	435	1	1	401	401
395	276	1	430	430	1	420	420	420	420	420	420	420	420	1	435	1	1	401	401		
396	276	276	1	420	420	420	1	1	1	1	420	1	1	435	435	435	435	435	1		
397	276	1	420	420	420	1	1	440	440	440	440	1	1	435	435	435	435	435	1	401	
398	1	420	420	420	1	1	440	440	440	440	440	1	435	1	1	1	435	435	435	1	401
399	1	420	420	420	1	1	440	440	440	440	440	1	435	1	1	1	435	435	435	1	401
400	1	420	420	1	440	440	440	440	440	440	1	1	447	447	447	447	447	1	435	1	401
401	1	420	1	1	440	440	440	440	440	440	1	1	447	447	447	447	447	1	401	401	401
402	276	1	440	440	440	440	440	1	451	451	1	447	447	447	447	1	401	401	401	401	401
403	276	1	440	440	440	440	440	1	451	451	451	451	1	447	447	447	1	401	401	401	401
404	276	1	440	440	440	1	451	451	451	451	451	1	276	1	1	401	401	401	401	401	401
405	276	1	440	1	451	451	451	451	451	451	1	276	276	276	1	401	401	401	401	401	401
406	276	276	1	451	451	451	451	451	451	451	1	276	276	276	276	1	401	401	401	401	401
407	276	276	1	451	451	451	451	451	451	451	1	276	276	276	276	1	401	401	401	401	401
408	276	1	451	451	451	451	451	451	451	451	1	276	276	276	276	1	401	401	401	401	401
409	276	1	451	451	451	451	451	451	451	1	276	276	276	276	276	1	401	401	401	401	401
410	276	1	451	451	451	451	451	451	451	1	276	276	276	276	276	1	401	401	401	401	401
411	276	1	451	451	451	451	451	451	451	1	276	276	276	276	276	1	401	401	401	401	401
412	276	1	451	451	451	451	451	451	451	1	276	276	276	276	276	1	401	401	401	401	401
413	276	276	1	451	451	451	451	451	451	1	276	276	276	276	276	276	1	401	401	401	401
414	276	276	1	451	451	451	451	451	451	1	276	276	276	276	276	276	1	401	401	401	401
415	276	276	1	451	451	451	451	451	451	1	276	276	276	276	276	276	1	401	401	401	401
416	276	276	1	451	451	451	451	451	451	1	276	276	276	276	276	276	276	1	401	401	401
417	276	276	1	451	451	451	451	451	451	1	276	276	276	276	276	276	276	276	1	401	401
418	276	276	1	451	451	451	451	451	451	1	276	276	276	276	276	276	276	276	276	1	1
419	276	276	276	1	451	451	451	451	451	1	276	276	276	276	276	276	276	276	276	276	276
420	276	276	276	1	451	451	451	451	451	1	276	276	276	276	276	276	276	276	276	276	276
421	276	276	276	1	451	451	451	451	451	1	276	276	276	276	276	276	276	276	276	276	276
422	276	276	276	1	451	451	451	451	451	1	276	276	276	276	276	276	276	276	276	276	276
423	276	276	276	1	451	451	451	451	451	1	276	276	276	276	276	276	276	276	276	276	276
424	276	276	276	276	1	451	451	451	451	1	276	276	276	276	276	276	276	276	276	276	276
425	276	276	276	276	1	451	451	451	451	1	276	276	276	276	276	276	276	276	276	276	276
426	276	276	276	276	276	1	451	451	451	1	276	276	276	276	276	276	276	276	276	276	276
427	276	276	276	276	276	276	1	1	276	276	276	276	276	276	276	276	276	276	276	276	276
	502	503	504	505	506	507	508	509	510	511	512	513	514	515	516	517	518	519	520	521	522

Figure 6.11. A 21 × 45-pixel portion of the image of geological contours in component-labeled form. Boundary pixels are represented by 1's, and values 2 and larger represent the sequential labels uniquely assigned to each area enclosed by the 1's.

12. The enclosed-area labeling has assigned a unique number to each area, but the map contains only a certain number of map units (phases). The online operator now assigns a unique serial number to each phase; that is, each enclosed area is identified as belonging to a given phase. The process is semiautomatic, carried out interactively, and the computer associates the phase labels with the area labels. During this process, missing phase labels and duplicate labels are detected. These errors are corrected (see Fig. 6.12) by displaying the appropriate areas and deleting or adding the labels online. It is necessary to return to earlier stages only if a topological error, such as a missing contour line, has been discovered. Additional software is desirable to facilitate detection of inconsistencies and operator errors at the earliest stages of processing.

13. The phase-labeled image may now be used in variety of ways, for example, to create a binary image for each phase, to study relationships between phases, and so on. The data which are in raster format (see Fig. 6.13), are stored on tape or disk for further analysis.

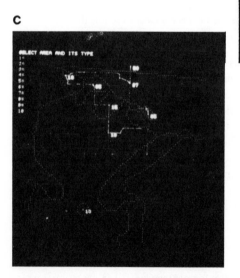

Figure 6.12. Interactive phase labeling of labeled areas from binary thinned contours: (a) the menu of options; (b) labeling areas; (c) eliminating a duplicate label.

	502	503	504	505	506	507	508	509	510	511	512	513	514	515	516	517	518	519	520	521	522
383	11	11	11	11	11	11	11	11	11	11	11	1	8	8	8	8	8	8	1	5	5
384	11	11	11	11	11	11	11	11	11	1	1	8	8	8	8	8	8	1	5	5	5
385	11	11	11	11	11	11	11	11	1	8	8	8	8	8	8	8	8	1	5	5	5
386	11	11	11	11	11	11	11	11	1	8	8	8	8	8	8	1	1	5	5	5	5
387	11	11	11	11	11	11	11	1	8	8	6	8	8	1	1	10	10	1	5	5	5
388	11	11	11	11	11	11	1	8	8	8	8	8	1	10	10	10	10	1	5	5	5
389	11	11	11	11	11	1	8	8	8	8	1	1	10	10	10	10	10	1	5	5	5
390	11	11	11	11	1	8	8	8	8	1	10	10	10	10	10	10	10	1	5	5	5
391	11	11	11	11	1	8	8	8	1	10	10	10	10	10	10	10	10	1	5	5	1
392	11	11	11	1	3	1	8	1	10	10	10	10	10	10	10	10	10	1	5	5	1
393	11	11	1	3	3	3	1	10	10	10	10	10	10	10	10	10	1	5	5	5	1
394	11	11	1	3	3	1	10	10	10	10	10	10	10	10	10	1	11	1	1	5	5
395	11	1	3	3	1	10	10	10	10	10	10	10	10	10	1	11	11	11	11	1	5
396	11	11	1	1	10	10	10	1	1	1	1	10	1	1	11	11	11	11	11	11	1
397	11	1	10	10	10	10	1	3	3	3	3	1	11	11	11	11	11	11	11	1	5
398	1	10	10	10	1	1	3	3	3	3	3	1	11	1	1	1	11	11	11	1	5
399	1	10	10	1	3	3	3	3	3	3	3	3	1	12	12	12	1	11	1	5	5
400	1	10	10	1	3	3	3	3	3	3	1	1	12	12	12	12	12	1	5	5	5
401	1	10	1	3	3	3	3	3	3	1	10	10	1	12	12	12	1	5	5	5	5
402	11	1	3	3	3	3	3	1	1	10	10	1	12	12	12	12	1	5	5	5	5
403	11	1	3	3	3	3	1	10	10	10	10	10	1	12	12	1	5	5	5	5	5
404	11	1	3	3	3	1	10	10	10	10	10	1	11	1	1	5	5	5	5	5	5
405	11	1	3	1	1	10	10	10	10	10	1	11	11	11	1	5	5	5	5	5	5
406	11	11	1	10	10	10	10	10	10	1	11	11	11	11	11	1	5	5	5	5	5
407	11	11	1	10	10	10	10	10	10	1	11	11	11	11	11	1	5	5	5	5	5
408	11	1	10	10	10	10	10	10	10	1	11	11	11	11	11	1	5	5	5	5	5
409	11	1	10	10	10	10	10	10	1	11	11	11	11	11	11	1	5	5	5	5	5
410	11	1	10	10	10	10	10	10	1	11	11	11	11	11	11	1	5	5	5	5	5
411	11	1	10	10	10	10	10	10	1	11	11	11	11	11	11	1	5	5	5	5	5
412	11	1	10	10	10	10	10	10	1	11	11	11	11	11	11	11	1	5	5	5	5
413	11	11	1	10	10	10	10	10	1	11	11	11	11	11	11	11	1	5	5	5	5
414	11	11	1	10	10	10	10	10	1	11	11	11	11	11	11	11	1	5	5	5	5
415	11	11	1	10	10	10	10	10	1	11	11	11	11	11	11	11	11	1	5	5	5
416	11	11	1	10	10	10	10	10	1	11	11	11	11	11	11	11	11	11	1	5	5
417	11	11	1	10	10	10	10	10	1	11	11	11	11	11	11	11	11	11	11	1	1
418	11	11	1	10	10	10	10	10	1	11	11	11	11	11	11	11	11	11	11	11	11
419	11	11	11	1	10	10	10	10	1	11	11	11	11	11	11	11	11	11	11	11	11
420	11	11	11	1	10	10	10	10	1	11	11	11	11	11	11	11	11	11	11	11	11
421	11	11	11	1	10	10	10	10	1	11	11	11	11	11	11	11	11	11	11	11	11
422	11	11	11	1	10	10	10	10	1	11	11	11	11	11	11	11	11	11	11	11	11
423	11	11	11	1	10	10	10	10	1	11	11	11	11	11	11	11	11	11	11	11	11
424	11	11	11	11	1	10	10	10	1	11	11	11	11	11	11	11	11	11	11	11	11
425	11	11	11	11	1	10	10	10	1	11	11	11	11	11	11	11	11	11	11	11	11
426	11	11	11	11	11	1	10	10	1	11	11	11	11	11	11	11	11	11	11	11	11
427	11	11	11	11	11	11	1	1	11	11	11	11	11	11	11	11	11	11	11	11	11
	502	503	504	505	506	507	508	509	510	511	512	513	514	515	516	517	518	519	520	521	522

Figure 6.13. A phase-labeled 21 × 45-pixel portion of the same image as shown in Figure 6.11. The 1's in this printout correspond to boundary pixels. The numbers 2 and higher are the new phase labels assigned interactively.

ALTERNATIVE APPROACHES

Figure 6.14 summarizes the previously described processing steps and the corresponding output data. Subimage creation and the subsequent patching to create mosaics were dictated by equipment limitations—for instance, too small a tablet—even though to have the option of creating mosaics is an advantage, irrespective of tablet size.

Following are some alternative approaches to processing:

1. If the map is digitized on a larger tablet, subimage generation can be avoided. The processing steps in Figure 6.14 from step iv onward remain unchanged. It would be highly advisable, however, to carry out the phase labeling as a part of the digitzng procedures, thereby eliminating most of the work in step vii, interactive labeling. Nevertheless, step vii will be required for possible online correcting or editing.

2. The contour maps also may be raster-scanned to create or obtain a binary raster-formatted mosaic directly. For example, a binary subimage can be obtained directly by scanning a 35-mm transparency of the submap or of the entire map. The maps, however, should be registered adequately on the scanner to avoid subsequent resampling. Processing step iv and subsequent steps remain unchanged.

3. The binary subimage for insertion into the mosaic can be created directly from the row-digitized data (the vectors) without storing it as a separate file.

4. The interactive editing stages in steps iv and vii somewhat depend on the peculiarities of the hardware and programming philosophy of the Modcomp II system available (see Appendix A). Some modifications to these programs would be needed before they can run successfully elsewhere.

CONCLUDING REMARKS

A variety of digitizing methods can be used with geological, geophysical, and geochemical maps and with cartographic data in general. Usually, to be practical, they are tailored to specific needs; see, for example, Peuker (1972), Davis and McCullagh (1975), Bryant and Zobrist (1976), Mitchell et al. (1977), and Bonfatti and Tiberio (1978). The basic principles of data capture and integation remain the same, however.

The main characteristics of the approach proposed in this chapter are the following:

1. that integration and processing are performed exclusively for data in raster form;

2. that the entire process for digitizing, preprocessing, and subsequent processing can be done on a minicomputer by a single individual; and

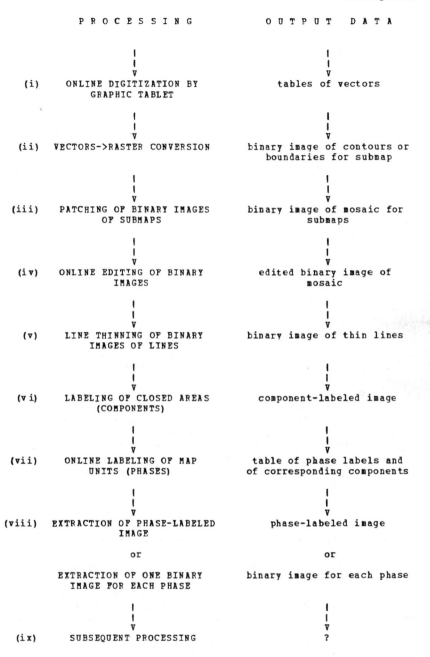

Figure 6.14. The procedure for processing maps.

3. that the process can be inexpensive because the computer is small, but only a few maps can be handled at the same time, unlike a process oriented toward production.

REFERENCES

Agterberg, F. P., C. F. Chung, A. G. Fabbri, A. M. Kelly, and J. S. Springer, 1972, Geomathematical Evaluation of Copper and Zinc Potential of the Abitibi Area, Ontario and Quebec, *Geol. Surv. Can., Paper 71-41*, 55p.

Anuta, P. E., H. Hauska, and D. W. Levandowski, 1976, *Analysis of Geophysical Remote Sensing Data Using Multivariate Pattern Recognition Techniques*, in Proc. Symposium on Machine Processing of Remotely Sensed Data, June 25-July 1, 1976, Purdue University, West Lafayette, Ind., Inst. Electr. Electron Eng., pp. 1B-11, 1B-14.

Bonfatti, F., and Tiberio, P., 1978, Data Management for Thematic Map Generation: *Comput. Graphics*, 71-78.

Bouille, F., 1976, Graph Theory and Digitization of Geological Maps, *J. Math. Geol.* **8:** 375-393.

Bryant, A. N., and A. L. Zobrist, 1976, IBIS: A Geographic Information System Based on Digital Image Processing and Image Raster Data Type, in Proc. Symposium on Machine Processing of Remotely Sensed Data, Purdue University, West Lafayette, Ind., June 29-July 1, 1976, pp. 1A.1-1A.7.

Davis, J. C., and M. J. McCullagh, eds., 1975, *Display and Analysis of Spatial Data*, Wiley, New York, 378p.

Fabbri, A. G., 1975, Design and Structure of Geological Data Banks for Regional-mineral Potential Evaluation, *Can. Inst. Mining Metall. Bull.* **68:**91-98.

Fabbri, A. G., S. R. Divi, and A. S. Wong, 1975, A Data Based for Mineral Potential Estimation in the Appalachian Region of Canada, in Report of Activities, *Geol. Surv. Can., Paper 75-1C* Part C, pp. 123-132.

Kasvand, T., A. G. Fabbri, and L. D. Nel, 1981, *Digitization and Processing of Large Regional Geological Maps*, Nat. Res. Council Can., Elec. Eng. Division, Report, ERB-938, 91p.

Leech, G. B., 1975, Project Appalachia, *Geol. Surv. Can., Paper 75-1*, Part C, pp. 121-122 (introduces 8 reports in ibidem, pp. 123-173).

Mitchell, W. B., S. C. Guttpill, K. E. Anderson, R. G. Fegeas, and C. A. Hallam, 1977, *GIRAS: A Geographic Information Retrieval and Analysis System for Handling Land Use and Land Cover Data*, U. S. Geological Survey, Prof. Paper 1059, 16p.

Peuker, T. K., 1972, *Computer Cartography*, Association of American Geographers, Resource Paper no. 17, Washington, D. C., 73p.

Robinson, J. E., 1982, *Computer Applications in Petroleum Geology*, Hutchinson Ross, Stroudsburg, Pa., 164p.

Sampson, R. J., 1975*a*, Surface II Graphic System, in *Geological Survey of Kansas*, J. C. Davis, ed., Series on Spatial Analysis no. 1, Lawrence, Kan., 240p.

Sampson, R. J., 1975*b*, The Surface II Graphic System, in *Display and Analysis of Spatial Data*, J. C. Davis and M. J. McCullach, eds., Wiley, London, pp. 244-266.

Srivastava, G. S., 1977, Optical Processing of Structural Contour Maps, *J. Math. Geol.* **9:** 3-38.

Switzer, P., 1976, Application of Random Process Models to the Description of Spatial Distributions of Qualitative Geologic Variables, in *Random Processes in Geology*, D. F. Merriam, ed., Springer-Verlag, New York, pp. 124-134.

CHAPTER 7

A Geological Database in Northwestern Manitoba, Canada

Chapters 7 and 8 apply image processing to the problem of integrating data from maps. A geological map is represented as a series of binary images, one image per stratigraphic unit. Contour maps are digitized as successive layers, one binary image per layer. Geological, geophysical, geochemical, structural, and mineral distribution maps, among others, can be assembled as a compatible database of black-and-white images. Overlay relationships, adjacency (contact) relationships, textural characteristics, and other features can be studied by means of techniques of image processing, thus enabling the testing of models and facilitating interpretation.

Whereas Chapter 8 is concerned with the analysis of such a database, this chapter covers the necessary geological background to the analysis. The geological description is made in terms of the binary images digitized.

THE DATA BASE

The map information digitized covers the Kasmere Lake-Whiskey Jack Lake (North Half) Area, in northwestern Manitoba, which is between longitudes 100°00' W and 102°00' W, and between latitudes 58°30' N and 60°00' N. Except for the gravity map, which has a scale of 1:500,000, all maps for the quantification were available in the UTM, projection, at a scale of 1:250,000, and cover a map area of approximately 20,000 km².

Figure 7.1 shows geological formations that represent the stratigraphic reconstruction described by Weber et al. (1975). The binary images in Figure 7.2

Pleistocene and Recent			Sand-gravel, boulder fields, clay (glacial till, eskers, fluvioglacial boulder deposits, lake-bottom sediments)	

GREAT UNCONFORMITY

P R E C A M B R I A N	P R O T E R O Z O I C A R C H E A N	A P H E B I A N	**Hudsonian igneous and metamorphic rocks** 22 Fluorite-bearing quartz nonzonite 21 Quartz-feldspar porphyry 20 Porphyritic quartz monzonite 19 Pink leucogranite to quartz monzonite, biotite-quartz monzonite 18 Migmetite: (a) metatexite; (b) diatexite 17 White granite to quartz monzonite, pegmatite, locally garnet-bearing

— INTRUSIVE CONTACT —

Metasedimentary rocks

13 Amphibolite, hornblendite 12 Meta-arkose 11 Conglomerate 10 Psammitic biotite gneiss, minor calc-sillicate rocks	16 Dolomite – Hurwitz Group 15 Argillite Group 14 Metagreywacke, metasiltstone Group

— DISCONFORMITY —

9 Albite-pyroxene rock
8a Calc-silicate rock
8b Marble
7 Pelitic biotite gneiss $\pm$ cordierite, garnet, sillimanite; minor calc-silicate rock and impure quartzite:
(a) porphyroblastic biotite gneiss; (b) augen gneiss;
(c) quartzite; (d) hypersthene paragneiss

— ? UNCONFORMITY ? —

6 Cataclastic biotite gneiss
5 Gray quartz dioritic to granodioritic gneiss: (a) amphibolite

— UNCONFORMITY —

ARCHEAN

Igneous rocks

4 Foliated quartz monzonite: (a) aplitic quartz monzonite
3 Foliated alaskite
2a Hypersthene-quartz diorite
2b Hypersthene trondhjemite
2c Hypersthene-quartz monzonite

— INTRUSIVE CONTACT —

1 Hypersthene gneisses

Figure 7.1. (*at left*). Geological formations in the study area. (*After W. Weber, D. C. P. Schledewitz, C. F. Lamb, and K. A. Thomas, 1975, Geology of the Kasmere Lake-Whiskey Jack Lake (North Half) Area (Kasmere Project), Manitoba Department of Mines, Resources and Environmental Management, Mineral Resources Division, Geological Services Branch, Pub. 74-2, p. 17.*)

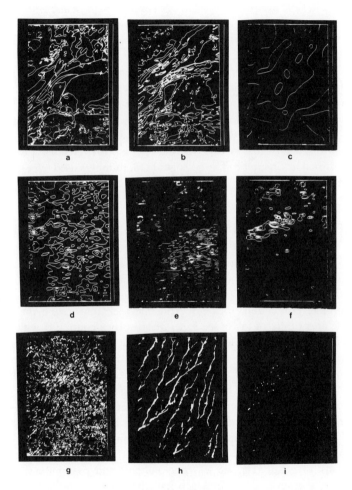

Figure 7.2. The multilayered data base in the Kasmere Lake-Whiskey Jack Lake area, northwestern Manitoba. (*a*) Bedrock geology boundaries; (*b*) aeromagnetic anomaly contours (gammas); (*c*) free-air Bougher gravity anomaly contours (milligals); (*d*) airborne gamma-ray spectrometric contours of the integral count of uranium equivalent concentration (eU), in ppm; (*e*) thorium equivalent (eTh), in ppm; (*f*) eU/eTh ratio; (*g*) topographic elevation contours (in 100-ft intervals); (*h*) the distribution of eskers and sand deposits; (*i*) 38 mineral occurrences located at the nearest pixel (167m).

represent a "multilayered" database of maps in registration (pixel-to-pixel correspondence).

The main source of information about the geology of the study area is a report by Weber et al. (1975), which, besides containing several geological and ancillary maps, provides much detail about the region's stratigraphy and economic geology. Other relevant sources are a gravity map series by Gibb and McConnell (1969) covering the area with a sparse set of measuring stations at 1:500,000 (Map 76: Wollaston Lake) and a set of three 1:250,000 contour maps of airborne gammaray spectrometric integer counts of equivalent uranium and of equivalent thorium (eU and eTh) and the eU/eTh ratio, made available by the Geological Survey of Canada (GSC, 1975).

Table 7.1 provides contour levels and background information, and Table 7.2 summarizes data on the 38 mineral occurrences located in Figure 7.1*I*. Occurrence types correspond to the groupings made by Weber et al. (1975).

Figures 7.3 to 7.5 provide examples of binary patterns extracted from the labeled images derived from the boundary images in Figure 7.2. The summary of the geology of the area that follows (see Weber et al., 1975) relates to the experiments described in Chapter 8. The geological unit numbers used refer to those in Table 7.2.

GEOLOGICAL SUMMARY

Archean foliated granitoid rocks (units 2 to 4) ranging from quartz diorite to alaskite granite occupy the southern half of the area, as shown in Figure 7.3A. Some of these rocks contain hypersthene and are associated to hypersthene gneisses (unit 1), probably the country rock. The Aphebian rocks that occur in the northern part of the area (units 14 to 16; see Fig. 7.3E) display a continuity with Hurwitz Group sedimentary units, while the remainder of the Aphebian sediments (units 7 to 13) at the center and in the southern part of the area lie in the extension of the Wollaston fold belt. Units 7, 8a and 8b, and 9 to 12 are shown in Figures 7.3B, 7.3C, and 7.3D, respectively. As can be seen by comparing Figures 7.3B to 7.3D with the pattern in Figure 7.4A, a central part of magnetic lows corresponds to the predominantly pelitic metasediments, surrounded by aeromagnetic highs (Figs. 7.4B and 7.4C) where Archean and Aphebian granitic and arkosic rocks occur, as shown in Figures 7.3D and 7.3F-H. Units 7 to 13, Aphebian metasedimentary rocks, consist of pelitic (Fig. 7.3 B), conglomeratic and psammitic (Fig. 7.3D), and calcareous strata (Fig. 7.3C), which have been interpreted as geosynclinal, platformal, and continental sediments. Younger igneous and metamorphic rocks generated during Hudsonian orogeny consist of migmatites, plugs, and stocks of anatexite and syn- and late-orogenic batholiths, most of which are massive and may truncate Hudsonian trends (units 17 and 19, Figs. 7.3F and 7.3G, and units 20 and 22, Fig. 7.3H). Figure 7.3*I* shows Pleistocene and recent sediments (drift) and water-covered areas where geological mapping was not available.

TABLE 7.1
Maps and Map Units of Ancillary Data Digitized

Map Number	Map Name	Map Description	Units Digitized
76	Wollaston Lake	1:500,000 Bouguer gravity anomaly contour map: contours at intervals of 5 milligals between −80 and −50	< −75, −70/−75, −65/−70, −60/−65, −55/−60, −50/−55
74-2-26	Kasmere Lake-Whiskey Jack Lake (North Half)	1:250,000 aeromagnetic anomaly contour map: contour values above 58,800 gammas ranging from 1800 to 3500	<1900, 1900/2100, 2100/2500, 2500/2900, 2900/3400, >3400
74-2-27	same as above	1:250,000 surficial geology map: patterns of esker deposits and of associated sand deposits	areas covered by either deposits
64N & 64K (North Half)	same as above	Topographic relief in 100-feet contour intervals between 900 and 1660	<1000, 1000/1100, 1100/1200, 1200/1300, 1300/1400, 1400/1500, >1500
eU ppm	same as above	Equivalent uranium in ppm; contour intervals in 0.2 ppm, between 0.0 and 6.0	0.0/1.0, 1.0/2.0, 2.0/3.0, 3.0/4.0, >4.0
eTh ppm	same as above	Equivalent thorium in ppm; contour intervals in 1.0 ppm, between 0.0 and 41.0	<10.0, 10.0/15.0, 15.0/20.0, 20.0/25.0, >25.0
eU/eTh	same as above	Ratio of equivalent uranium to equivalent thorium contour intervals in 0.01 units between 0.0 and 0.55	<0.20, 0.20/0.25, 0.25/0.30, 0.30/0.35, 0.35/0.50, >0.50
74-2-30	same as above	Economic geology: location of the first 32 mineral occurrences in Table 7.2, for the commodities pyrite, pyrrohite, sulfide, copper, polybdenum, cobalt, nickel, zinc, lead, uranium, and gossan	32 pairs of $x-y$ coordinates corresponding to the location of the nearest 167 × 167-m pixels
64N & 67K (North Half)	CANMINDEX computer plot	1:250,000 Calcomp plot: location of 12 uranium occurrences, 6 of which coincide with location on map 74-2-30 and the last 6 of Table 7.2	6 pairs of $x-y$ coordinate, as above

In apparent coincidence with the Wollaston fold belt, there is a weak linear belt of low gravity values, containing a chain of isolated "lows" that occur within an area of gradually increasing gravity values, as shown in Figures 7.4D through 7.4F. According to Weber et al. (1975), the "low" over the pelitic sediments within the

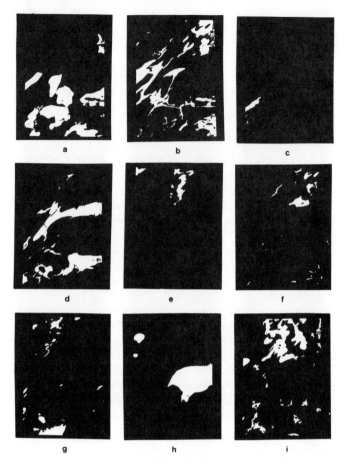

Figure 7.3. Some geological binary images obtained by processing the bedrock geology boundary image. *(a)* Archean igneous rocks; *(b)* Proterozoic or Aphebian pelitic metasediments (biotite gneiss); *(c)* Aphebian calc-silicate rocks; *(d)* Aphebian psammitic and conglomeratic metasediments (biotite gneiss, conglomerate, and metaarkose; *(e)* Aphebian units of the Hurwitz Group (dolomite, argillite, meta greywacke, and metasiltstone); *(f)* white granite to quartz monzonite and pegmatites; *(g)* pink leucogranite to quartz monzonite; *(h)* porphyritic and fluorite-bearing quartz monzonite; *(i)* Pleistocene, Recent, and unmapped areas.

Wollaston fold belt could be a result of widespread partial melting that formed large bodies, which, besides being exposed at the surface, are much larger at depth. As suggested for the Hurwitz Group, such a pattern would indicate a cratonic basin.

The rocks in the area underwent regional metamorphism of upper greenschist to lower amphibolite facies in the northwestern part, to upper amphibolite-hornblende granulite facies in the southwest (Schledewitz, 1978).

Such data, however, have not been digitized in this study. Lithostratigraphic

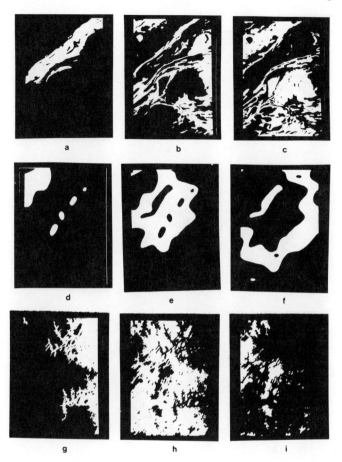

Figure 7.4. Some geophysical interval and topographic elevation binary images obtained by processing geophysical anomaly contour interval images: (*a*) aeromagnetic anomaly lows <2100 gammas; (*b*) aeromagnetic anomaly interval between 2100 and 2500 gammas; (*c*) aeromagnetic anomaly interval between 2500 and 2900 gammas; (*d*) gravity lows < −75 milligals; (*e*) gravity lows between −75 and −70 milligals; (*f*) gravity interval between −65 and −70 milligals; (*g*) topographic elevation below 1100 feet; (*h*) topographic elevation between 1100 and 1300 feet; (*i*) topographic elevation above 1300 feet.

units such as Aphebian pelitic biotite gneisses (unit 7, Fig. 7.3*B*) and psammitic gneisses, conglomerates, and meta-arkoses (units 10 to 12, Fig. 7.3*D*) define large-scale folds. The two main trends in the area are indicated by major and minor folds and their related planar and linear structures. The major tectonic trend is in the northeastern direction: it occurred during Aphebian time and is represented typically by the Wollaston fold belt. A pre-Aphebian, less pronounced and slightly curvilinear fold trend occurs in a roughly eastern direction. A third minor folding

93

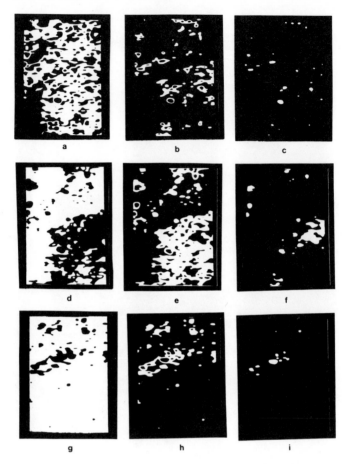

Figure 7.5. Some binary images obtained by processing the radiometric contour binary maps: (a) eU < 2 ppm; (b) eU between 2 and 3 ppm; (c) eU > 3 ppm; (d) eTh < 2 ppm; (e) eTh between 2 and 3 ppm; (f) eTh > 3ppm; (g) eU/eTh < 0.20; (h) eU/eTh between 0.20 and 0.30; (i) eU/eTh > 0.30.

phase has also occurred in the area. Structural maps and tectonic maps that also accompany the report by Weber et al. (1975) have not been digitized.

Three-quarters of the area is covered by glacial drift, whereas the rest is mostly covered by lakes and rivers, leaving less than 2 percent to rock exposures. Till, sand, and gravel associated with the esker system make up the drift cover. The till is composed of sand, gravel, and large, rounded boulders up to 2 m in diameter. These may form extensive boulder fields in low-lying areas. Some of the more resistant granitic boulders can be traced from local bedrock outcrop; in most cases the boulders are not transported far from their sources. Many eskers, which form southwesterly extending long, sinuous ridges, occupy topographic depressions typical of meltwater channels The ridges run subparallel and are spaced approxi-

94

TABLE 7.2
Summary Data on Mineral Occurrences in the Study Area

Number	Occurrence Type[a]	Commodities Present[b]	Picture Coordinates of Closest Pixel	
			x	y
1	1, 4	U, Mo, Cu, Ni	150	111
2	3, 2	Pb, Cu, Ni	441	160
3	2	Cu, Pb, Zn, Ni	292	235
4	2	Cu, Pb, Ni	260	250
5	5, 2	G	257	254
6	3, 2	Pb, Cu, Ni, Co?	241	286
7	1	U	299	315
8	3, 2	Co, Ni, Cu, Pb, Zn	264	383
9	3, 2	Co, Ni, Cu, Pb, Zn	210	384
10	2	Cu, Zn, Pb, Ni, Co	168	398
11	1, 2, 3	U, Ni, Cu, Co, Pb, Zn, Th	125	418
12	3, 2	Co, Ni, Cu, Pb, Zn	257	436
13	1	U	261	438
14	6	Po, Py	108	439
15	6	Py	220	449
16	6, 2	Py	134	460
17	2	Zn, Cu	137	547
18	6	Py	94	561
19	2	Zn, Cu	62	592
20	2	Zn, Cu	58	595
21	2	Zn, Cu	53	599
22	2	Zn, Cu	56	600
23	2	Zn, Cu	51	604
24	4, 2	Mo, Cu, Ni	248	636
25	1	U	147	244
26	2, 1	Cu	101	860
27	2	Cu	91	940
28	2	Cu	100	937
29	4, 2	Cu, Mo	486	863
30	2	Zn, Cu, Pb	637	754
31	5, 2	G	642	699
32	6, 2	S	703	666
33	1	U, Th, Mo	246	386
34	1	U, Th	160	381
35	1	U, Th	178	370
36	1	U	516	139
37	1	U	314	340
38	1	Th, U	365	295

[a] Types: 1 = U, 2 = Cu, Cu−Zn, 3 = Co−Ni, 4 = Mo, 5 = G, 6 = Py, Po,S
[b] Commodities: Cu = copper, Zn = zinc, Co = cobalt, Ni = nickel, U = uranium, Mo = molybdenum, G = gossan, Py = pyrite, Po = pyrrothite, S = sulphur

mately 15 km apart. They meander in a southerly to southwesterly direction and form steep-sided ridges, which rise up to 50 m above the surrounding drift. Their pattern can be seen in Figure 7.2H.

Elevations in the area are between 300 and 600 m above sea level. The percentage of bedrock exposure is lowest in a central belt in the area (see Fig. 7.4G), where the poor bedrock exposure corresponds to more easily eroded pelitic metasedimentary rocks. The digitized map pattern, however, does not show this relationship. In the southwestern and northwestern parts of the area, more resistant granitic and arkosic rocks tend to form hills of up to 100 m relief (see Figs. 7.4H and 7.4I).

Some of the patterns extracted from the airborne gammaray spectrometric contour maps of Figures 7.2D to 7.2F are shown in Figure 7.5. The contour values for the radioelement concentrations are "average surface concentrations" over the area sampled by the airborne spectrometer, which include some outcrop, overburden, and water in small ponds, streams, and swamps, providing contour values that are considerably lower than the concentrations in the bedrock but that still reflect the regional distribution of the elements in it. Because of compilation and contouring procedures, the contours in some cases may be distorted in the east-west direction.

EXPLORATION AND MODELS OF MINERALIZATION

In 1967 several companies began to conduct combined magnetic and radiometric surveys along the main metasedimentary part of the Wollaston fold belt. A first discovery of uranium mineralization at Rabbit Lake, near Wollaston Lake, caused an increase in uranium and base metal exploration activity all along the metasedimentary belt in the Kasmere Lake area, which by 1969 was covered by exploration reservations. Airborne surveys were followed by electromagnetic, geological, and geochemical mapping. In spite of this, only a few radioactive and magnetic anomalies were drilled because of the very small outcrop areas. No major discovery was found, and by 1975 all exploratory reservations and claim blocks had lapsed. Regional geochemical surveys, carried out in northwestern Manitoba during 1975, have been described by Coker (1976). As mentioned by Soonavala, Garber, and Whitworth (1979), new exploration activity started in 1976 and brought the staking of all the areas characterized by anomalously high uranium. More recent exploration activities in the area consisted of additional mapping, and more detailed airborne spectrometric surveys as part of a Federal/Provincial Uranium Reconnaissance Program.

Almost all mineral occurrences were discovered by means of airborne geophysical methods. In general, the radioactive anomalies in the area (see Figs. 7.5A to 7.5C) are caused by boulder fields of slightly radioactive granitic rocks, containing disperse uranium-bearing minerals or their weathering products. The intensities of the anomalies are exaggerated because of the proportionally greater surface area exposed in the boulder fields.

Weber et al. (1975) describe disseminated uranium mineralization in:

I. white granite and pegmatite (unit 17) and similar granitic rocks occurring as sills in pelitic biotite gneisses (unit 7);

II. red and pink pegmatite (part of unit 19);

III. calc-silicate gneiss (unit 8a) and marble (unit 8b);

IV. foliated quartz-monzonite (unit 4); and

V. granitic and arkosic granite gneiss or meta-arkose (units 19 and 12).

They conclude that uranium mineralization is restricted to units 17, 19, 8a, 12, and 4 (in order of importance).

Occurrences of base-metal mineralizations are grouped into the following classes:

1. Cobalt-nickel in metasedimentary rocks of the Wollaston fold belt, for example, within calc-silicate rocks (unit 8a) and marble (unit 8b) near the contact with meta-arkose (unit 12);

2. zinc-copper (and copper-lead-zinc), in (a) zones of massive and disseminated sulphides in calc-silicate rocks (units 8a and 8b) near the contact with hypersthene-quartz monzonite (unit 2c), (b) disseminated sulfides in pelitic gneisses (unit 7), and (c) as Cu-Pb-Zn in pelitic gneiss; and

3. copper (with or without molybdenum) as sparsely disseminated mineralization in the southern part of the area in pelitic gneisses (units 7 and 7d) in calc-silicate interlayers (units 8a and 8b) and in pegmatitic calc-silicate rocks.

CONCLUDING REMARKS

The geological and ancillary information described here is used in part to model experiments for relating 12 uranium occurrences (groups I to IV) and 24 base metal occurrences (group II) to geological and ancillary map patterns in their vicinities. The applications presented in Chapter 8 use transformations of binary compressed images and concepts of mathematical morphology developed by Matheron (1975) and Serra (1976) in France. The binary images are transformed and combined by means of logical operations to obtain derived patterns as coincidences to desirable characteristics. The proportions of pixels in these patterns over the total number of pixels in the study area can be considered as the probability that a random pixel swept throughout the image "hits" the pattern. Patterns of probabilities can be built, displayed, and quantitatively characterized for regional mineral resource evaluations. Thus the technique described here can be considered a new geological tool.

REFERENCES

Coker, N. B., 1976, Geochemical Follow-up Studies, Northwestern Manitoba, *Geol. Surv. Can., Paper 76-1C*, pp. 263-267.

Gibb, R. A., and R. K. McConnell, 1969, The Gravity Anomaly Field in Northern Manitoba and Northeastern Saskatchewan with Maps No. 68 to 76, in *Gravity Map Series of the Dominion Observatory*, Ottawa, Canada, 28p.

GSC, 1975, *Airborne Gamma-ray Spectrometric Survey*, Resources Geophysics and Geochemistry Division, *Geol. Surv. Can., Open Files 317 and 318*.

Matheron, G., 1975, *Random Sets and Integral Geometry*, Wiley, New York, 261p.

Schledewitz, D. C. P., 1978, Patterns of Regional Metamorphism in the Churchill Province of Manitoba (north of 58 degrees), in J. A. Fraser and W. W. Heywood, eds., *Metamorphism in the Canadian Shield*, Geol. Surv. Can., Paper 78-10, pp. 179-190.

Serra, J., 1976, *Lectures on Image Analysis by Mathematical Morphology*, Cahier N-475, Centre de Morphologie Mathematique, Fontainebleau, France, July 1976, 225p.

Soonawala, N. M., R. J. Garber, and R. A. Whitworth, 1979, Follow-up of the Uranium Reconnaissance Program in Northwest Manitoba, *Can. Inst. Min. Metall. Bull.* **72:** 83-94.

Weber, W., D. C. P. Schledewitz, C. F. Lamb, and K. A. Thomas, 1975, *Geology of the Kasmere Lake-Whiskey Jack Lake (North Half) Area (Kasmere Project)*, Manitoba Department of Mines, Resources and Environmental Management, Mineral Resources Division, Geological Services Branch, Pub. 74-2, 163p.

8

Quantitative Characterization of Geological and Ancillary Map Patterns in Northwestern Manitoba

Drawing on the mineralization concepts of Chapter 7, this chapter uses a number of separate models that were built for all lithostratigraphic units considered to be important hosts for uranium and for the base metals. The applications performed here use transformations of binary images. Logical operations or Boolean algebra, and binary neighborhood transformations are computed on a minicomputer. This approach has been proposed earlier, by Fabbri and Kasvand (1981) and by Fabbri (1981).

Figure 8.1 shows examples of such operations and transformations. In Figure 8.1A, D is the pattern of 61 × 61 pixel neighborhoods (corresponding to 10 × 10-km squares) of 12 uranium occurrences; it can be generated by 30 successive dilatations of the original image by a 3 × 3-pixel black template. On that image the 12 occurrences are identified by 12 black pixels in point-to-point correspondence with the occurrence locations. Each black pixel grows into an array of 61 × 61 black pixels. A measure of D is its area, mes $D = 38,973$ pixels, each corresponding to a square area of 167 m × 167 m. We can say also that 0.057, or 5.7%, is the areal proportion of the pattern, that is, the probability that a 10 × 10-km square array of pixels "hits" or contains an occurrence of uranium. Incidentally, 2^{-5} or 0.002%, is the probability that a single pixel translated at random throughout the study area of 684,076 pixels "hits" a uranium occurrence pixel. (The total image area, 760 × 1,004 (= 763,040) pixels, is 78,964 pixels larger than the study area within the geological boundary image.) The area of the pattern G, Aphebian pelitic metasediments (Fig. 8.1B), is 119,212 pixels; the probability that a random pixel "hits" G is 0.174, or 17.4%. The intersection

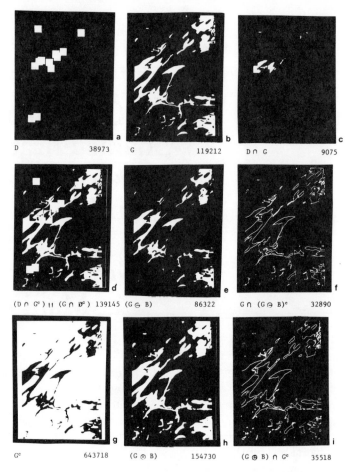

<table>
<tr><td>D</td><td>38973</td><td>G</td><td>119212</td><td>D ∩ G</td><td>9075</td></tr>
<tr><td>(D ∩ Gᶜ) ‖ (G ∩ Dᶜ) 139145</td><td></td><td>(G ⊖ B)</td><td>86322</td><td>G ∩ (G ⊖ B)ᶜ</td><td>32890</td></tr>
<tr><td>Gᶜ</td><td>643718</td><td>(G ⊕ B)</td><td>154730</td><td>(G ⊕ B) ∩ Gᶜ</td><td>35518</td></tr>
</table>

Figure 8.1. Transformations of binary images. The numbers of black pixels in the images (here displayed in white on a Tektronix 611 storage display unit) are shown below the right corners of the plots. The expressions for the transformations are shown below the left corners. (*a*) Image *D* of 61 × 61-pixel neighborhoods (10 × 10-km squares) of 12 uranium occurrences; (*b*) the image G of Aphebian pelitic metasediments; (*c*) the intersection (overlap or coincidence) between *D* and G; (*d*) the image produced by the .EXOR. (exclusive .OR.) logical operation between *D* and G, which shows one image in the context of the other; (*e*) image G eroded by a 5 × 5-pixel black template; (*f*) image of the pixels eroded from G; (*g*) the image of the complement of negation of G; (*h*) image G dilatated by a 5 × 5-black template; (*i*) image of the black pixels "added" to the image G during the dilatation.

or overlap — or better, the coincidence between the two patterns ($D \cap G$ in Fig. 8.1C) has an area of 9075 pixels; the probability that the random pixel hits both patterns in coincident positions is 0.013, or 1.13%. Figure 8.1D shows the pattern obtained by the union of the two nonoverlapping subsets of *D* and G; it represents another way of visually displaying their relationship. Figure 8.1E shows $G \ominus B$; that is, G eroded by B, a 5 × 5 pixel black template. Figure 8.1F shows the pattern of eroded pixels, Figure 8.1G the pattern of the complement of G, and Figure 8.1H the pattern of $G \oplus B$; that is, of G dilatated by B. The pattern of the black pixels "added" to G during the dilatation is shown in Figure 8.1I.

DERIVATION OF BINARY PATTERNS RELATED TO URANIUM MINERALIZATION

Binary images are transformed and combined to obtain derived patterns as coincidences of desirable characteristics. Let us consider an application of such an approach. All the zones within 420 m of the gradational contact between Aphebian pelitic metasediments (units 7 and 7c in Fig. 7.1) and the more porous conglomeratic and psammitic Aphebian metasediments (units 9, 10, 11, and 12), which are considered potential traps to uranium mineralization due to the good porosity, have been extracted as follows. First the union between the pattern (7) ∪ (7c) and the patterns (9) ∪ (10) ∪ (11) ∪ (12) were computed (see Figs. 7.3B and C in chapter 7). The resulting binary images were then dilatated by a square structuring element of 5 × 5 black pixels, and the intersection between the two dilatated patterns was computed. This represents the 840-m-wide zone, which in Figure 8.2A is displayed together with the pattern of 10 × 10-km cells centered around the 12 uranium occurrences. What is represented is the result of the logical operation of exclusive .OR., .EXOR., the union of nonoverlapping parts of two sets.

This artifice is used here to display with one binary image the relationship between two binary patterns characterized by strongly different shapes. Here a linear pattern and a pattern of squares are clearly visible, and the interruption of the line pattern, shown as the holes in the square pattern, represents the overlap or coincidence between them. We can see in Figure 8.2A, the gradational contact between Aphebian metasediments of different porosity: it coincides with part of the neighborhoods around the uranium occurrences. The same artifice, which allows us to see one pattern in the context of another, is used in the remainder of the binary images shown in Figure 8.2.

Model 1

The map patterns have been used to relate the overlaps of features extracted from them with uranium mineralization, according to a model that considers the following:

1. The geological map shows a promising contact between pelitic and psammitic rocks. Marble and calc-silicate rocks occur along this contact.

101

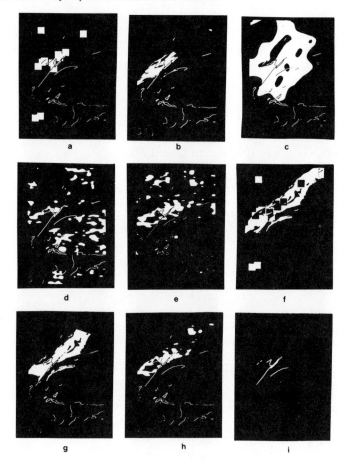

Figure 8.2. Partitioning of the binary image of a gradational contact into environments favorable to uranium mineralization. Part *a* shows a binary image of the 50-m-wide contact zone between Aphebian pelitic metasediments and the coarser psammitic and conglomeratic metasediments of Aphebian age. The contact image is compared (by .EXOR.'ing) with the square areas surrounding the 12 uranium occurrences; with the image of aeromagnetic lows < 1900 gammas (*b*); with the image of gravity lows < −70 milligals (*c*); the images of eU highs > 2 ppm (*d*); and with the image of eU/eTh ratio highs > 0.20. (*e*) In part *f*, the image of aeromagnetic anomaly lows < 2100 gammas is compared with the image of the square areas surrounding the 12 uranium occurrences; part *g* shows promising parts of the image of the contact that coincide with areas in which gravity lows (< −70 milligals) and aeromagnetic lows (< 2100 gammas) coincide, that is, the line pattern cutting across the larger shapes. Part *h* shows promising parts of the image of the contact that coincide with areas in which aeromagnetic lows (< 2100 gammas) and eU/eTh ratio highs (> 0.20) coincide, that is, the line pattern cutting across the larger shapes. Finally, part *i* depicts the union of the intersections between the contact, aeromagnetic and gravity lows, and the contact, aeromagnetic lows, and eU/eTh highs, that is, the extracted pattern.

2. Gravity and aeromagnetic anomaly lows occur, which may relate to areas where the sedimentary belt is thickest.

3. The occurrence of eU/eTh highs in these areas seem to relate to the development of white anatectic granite and pegmatite within the pelitic rocks.

As exemplified in Figure 8.2, the feature-extraction experiment that follows considers coincidences of the pattern of gradational contacts with other binary patterns, such as the pattern of aeromagnetic anomaly lows < 900 gammas (Fig. 8.2B), the pattern of gravity lows less than −70 milligals (Fig. 8.2C), the pattern of eU highs > 2 ppm (Fig. 8.2D), and the pattern of eU/eTh highs > 0.20 (Fig. 8.2E).

In Figure 8.2F, a comparison is made between the pattern of aeromagnetic lows < 2100 gammas (which corresponds to the metasediments of the Wollaston fold belt) and the pattern of 10 × 10-km neighborhoods around the 12 uranium occurrences. Figure 8.2G shows the relationships between the Aphebian gradational contact, the aeromagnetic lows < 2100 gammas, and the gravity lows < −70 milligals. Here the coincidence or areal proportion is 0.005 or 0.5 percent, which can be compared to 1.9 percent of the entire gradational contact in the study area. In Figure 8.2H, the coincidence is shown between the gradational contact, the pattern of pixels corresponding to the aeromagnetic lows < 2100 gammas, and eU/eTh highs > 0.2. This coincidence corresponds to 0.2 percent of areal proportion, and is almost entirely a subset of the pattern in Figure 8.2G. This can also be seen by computing the union of the two patterns, displayed in Figure 8.2I. This pattern represents a probability of a 0.5 percent that a random pixel hits a pixel belonging to our Aphebian gradational contact, where it either coincides with aeromagnetic lows and gravity lows or coincides with aeromagnetic lows and eU/eTh highs.

Expressions for the operations and transformations can be written as follows. If mnemonics are used for indicating the binary images, we can write G1 and G2 for the images of the two Aphebian metasedimentary units, AL for aeromagnetic anomaly lows, GL for gravity anomaly lows, and UT for eU/eTh ratio highs. Then the 850-m-wide gradational-contact-zone binary image− set CT between map units G1 and G2−can be written as follows:

$$CT = (G1 \oplus B) \cap (G2 \oplus B)$$

The extracted pattern, shown in Figure 8.2I, can be written as

$$EP = (AL \cap GL \cap CT) \cup (AL \cap UT \cap CT).$$

This aspect of the uranium depositional environment in the area can be considered a probability; furthermore, it can be characterized in several different ways either by itself, or in the context of other patterns.

The results of these experiments point to the portions of the Aphebian metasedimentary gradational contact within the Wollaston fold belt where uranium occurrences have not been discovered yet. Additional information—such as more recent geological mapping, geochemistry, or a more detailed gammaray spectrometric survey—which has not been considered here might restrict these searching areas even further down to operational areas for direct exploration.

Model 2

Aspects of uranium-related map patterns within other lithologies can be extracted for a model of areas characterized by aeromagnetic lows (< 2100 gammas), and eU/eTh highs (> 0.20).

Figure 8.3A shows the coincidence between the pattern of eU/eTh highs and that of the uranium occurrence in 10 × 10-km neighborhoods. Figure 8.3B shows the coincidence between the latter and the Aphebian pelitic metasediments (unit 7). Figure 8.3C shows the pattern of Aphebian metasediments where both conditions mentioned in the preceding paragraph hold. The same two conditions are modeled for the patterns in Figures 8.3E and 8.3G.

Figure 8.3D shows the coincidence between map units 17 and 19, white and pink granites and pegmatites, respectively, and the uranium occurrence neighborhoods. This coincidence is poor, as can be expected, for heavily underrepresented map units such as the pegmatites, most of which cannot be properly displayed on either 1:250,000 or 1:50,000 maps. This is so because such rocks occur as outcrops that are too small to be drawn at those two scales. Nevertheless, as shown in Figure 8.3E, from the binary images of those map units, it is possible to extract the portions of terrains that are characterized by the two coincident conditions previously selected for Figure 8.3C. Similar considerations can be made for the calc-silicate rock units (8a and 8b), and their coincidence with the uranium occurrence neighborhoods in Figure 8.3F. Here only one outcrop can be extracted from the image of these terrains, as can be seen in Figure 8.3G, for those same conditions.

Model 3

The last lithological environment tentatively considered for uranium occurrences is in Archean igneous rocks (map units 1, 2a, 2b, 2c, 3, 4a, and 4 in Figure 7.3A). Weber et al. (1975) consider disseminated uranium mineralization in map unit 4 (see Chapter 7). In Figure 8.3H, this is shown in coincidence with the pattern of the uranium-occurrence neighborhoods. For the two occurrences to the southwest of the image, two broad characteristics seem to concide with these Archean igneous rock map units: aeromagnetic anomaly values ranging between

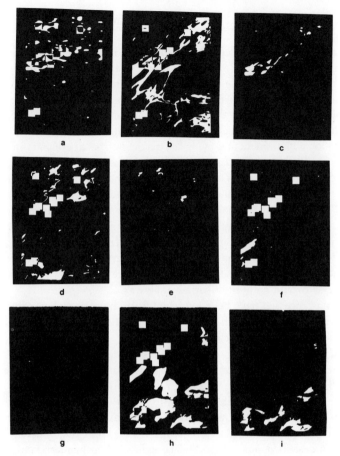

Figure 8.3. Other aspects of uranium-related environments. *(a)* comparison between eU/eTh ratio highs (> 0.20) and the square areas around the 12 uranium occurrences; *(b)* comparison between the image of Aphebian pelitic metasediments with the square areas around the uranium occurrences; *(c)* binary image of the areas in which Aphebian pelitic metasediments, aeromagnetic anomaly lows (< 2100 gammas), and eU/eTh highs (> 0.20) conincide; *(d)* comparison between the image of white and pink granites and pegmatites and that of the square areas around the 12 uranium occurrences; *(e)* binary image of the areas in which white and pink granites and pegmatites coincide with aeromagnetic anomaly lows (> 2100 gammas) and eU/eTh highs (> 0.20); *(f)* comparison between the image of Aphebian calc-silicate rocks and that of the square areas around the 12 uranium occurrences; *(g)* binary image of the areas in which Aphebian calc-silicate rocks coincide with aeromagnetic lows (< 100 gammas) and with eU/eTh ratio highs (> 0.20); *(h)* comparison between the image of Archean igneous rocks and that of the square areas around the uranium occurrences; *(i)* binary image of the areas in which Archean igneous rocks coincide with gravity highs (> −55 milligals) and aeromagnetic highs (> 2500 gammas).

2100 and 2900 gammas, and gravity anomaly values ranging between −55 and −60 milligals.

This pattern of coincidences, shown in Figure 8.3*I*, is arbitrary, and probably not interpretable without additional information. However, patterns like this last one are not easily imaginable and cannot be practically extracted by hand from sets of maps. They can induce the geologist to postulate models that should be more realistic in terms of what the image database contains. This may be particularly important in instances in which a significant lack of control exists for the regional mineralization environment and many fast trial-and-error experiments are required.

To conclude this section, it must be duly noted that the derived patterns described as subsets of pixels or regional environments can also be generated by other techniques of automatic classification, which are applied, for example, either in remote sensing (pixel classification) or in several multivariate statistical analysis applications. In many instances, however, a geologist may prefer to look at patterns of overlays such as the ones shown here, and observe metallogenic models developed as combinations of binary patterns at first, before considering automatically produced probability contours. Figure 8.4 shows two such contours, produced by two different multivariate analyses.

The data processed have been extracted from the phase-labeled images used in this section for producing the binary patterns. The seven phase-labeled images produced from the binary contour images (shown in Figs. 7.2A to 7.2G in chapter 7), the binary image of eskers and sand deposits (in Figure 7.2*H),* and the information on location in picture coordinates and on occurrence classification for the 38 occurrences (shown in Figure 7.2*I)* have been transferred to a different computer. The data have been entered into SIMSAG, an interactive system with graphical input/output for multivariate statistical analysis, developed at the Geological Survey of Canada by Chung (1979). From the integration of the information described, augmented by lake-sediment geochemical data, evaluation of mineral and energy resources has been performed by Chung (1983) and by Bonham-Carter and Chung (1983).

The contour maps in Figure 8.4 have been generated as a preliminary attempt to detect favorable uranium mineralization using the same overall models of the experiments for Figures 8.2 and 8.3. The results, shown in the two contour patterns, are very close to those in the binary patterns derived here. As expected, the aeromagnetic anomaly lows (< 2100 gammas) that characterize the Wollaston belt show up as the most important variable in the analysis. In the contour maps the high values represent environments similar to those surrounding the known occurrences of uranium. The contour values have been computed for square cells measuring approximately 9.5 km × 9.5 km (= 90 km^2). the data have been transformed into binary (0-1) variables indicating the absence or presence of interval values of geophysical anomaly or geological map-unit labels or mineral occurrences for each type of occurrence in each square cell.

Further experiments of this type are being considered as extensions of the approach. For example, the images of eskers and sand-deposit distribution and

106

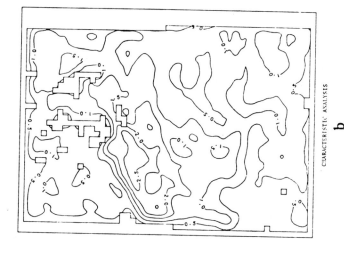

a

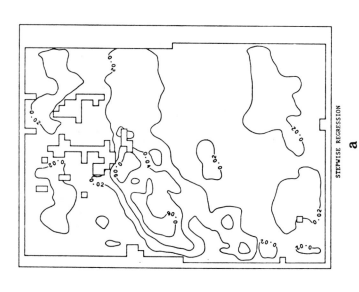

b

Figure 8.4. Probability contours obtained by analyzing the image data in the Kasmere Lake-Whiskey Jack Lake area by SIMSAG, an interactive package for statistical analysis programmed by Chung (1979). (a) The contour pattern obtained by stepwise regression; (b) the contour pattern obtained by characteristic analysis. The high contours indicate areas more favorable to uranium mineralization.

107

the contour interval of the topographic relief in Figures 7.2G and H may be used in order to correlate those binary patterns with the patterns for gammaray radiometric contours and those for some of the geological map units. Additional experiments of this type are beyond the scope of this section. Clearly, a complete study of the regional patterns related to uranium mineralization in the study area would be a major undertaking. What has been introduced here is a methodology to point out the kind of problems and results to be expected. A second application on different commodities, Cu-Pb-Zn mineralization, is described in the next section.

DERIVATION OF BINARY PATTERNS RELATED TO BASE METAL MINERALIZATION

Zinc-copper mineralization is described in this section in terms of binary patterns extracted from the phase-labeled images of bedrock geology, aeromagnetic anomaly, and gravity anomaly contour maps (shown in chapter 7 in Figs. 7.2A, B, and C, respectively. Zones of massive and disseminated sulfides in calc-silicate rocks (units 8a and 8b in Figure 8.5B), near the contact with hypersthene-quartz monzonite (unit 2c in Fig. 8.5A), disseminated sulfides in pelitic gneisses (unit 7 in Fig. 8.5C), and Cu-Pb-Zn in pelitic gneisses (unit 7) are correlated with the patterns of 10 × 10-km neighborhoods around 24 basemetal occurrences known in the area.

The intersection between the latter pattern of occurrences and the union of the binary patterns of map units 2c, 8a, 8b, and 7 represent approximately 39% of the area of the binary pattern of base metal occurrence. In the following derived patterns, this area of overlap is subdivided into portions that coincide with combinations of patterns for different ranges of aeromagnetic and gravity anomaly values. Of the many binary patterns so obtained, six have been selected that coincide best with the binary pattern of the 10 × 10-km neighborhoods of the Cu-Pb-Zn occurrences. Occupying the central part and also much of the area of the neighborhoods, they are inferred to represent environments similar to those in the neighborhoods.

The six derived patterns shown in Figures 8.5D to I are the following:

8.5D. Terrains for which aeromagnetic anomaly values are < 2100 and gravity anomaly values range between −65 and −70: three occurrences are in this environment.

8.5E. Terrains for which aeromagnetic anomaly values are < 2100 gammas, and the gravity anomaly values range between −60 and −65 milligals: three occurrences.

8.5F. Terrains for which the aeromagnetic anomaly values range between 2100 and 2500 gammas, and the gravity anomaly values are < −65 milligals: six occurrences.

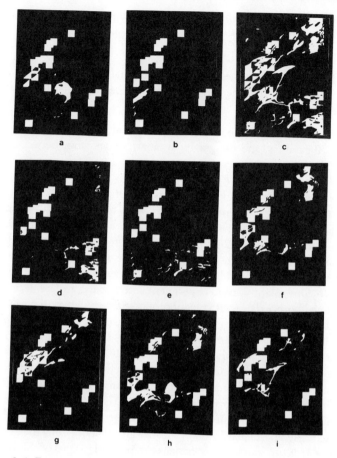

Figure 8.5. Extraction of binary images representing environments related to base metal mineralization in the Kasmere Lake-Whiskey Jack Lake area in northwestern Manitoba, showing (a) a comparison (by .EXOR.'ing) between the image of Aphebian hypersthene-quartz monzonite and the 10 × 10-km square areas around the 24 base metal occurrences, and comparisons of the image of base metal occurrences with the image of Aphebian calc-silicate rocks (b); with that of Aphebian pelitic metasediments (c); with that of the areas in which aeromagnetic anomaly values are < 2100 gammas and the gravity values range between −65 and −70 milligals (d); with that of the areas in which aeromagnetic anomaly values are < 2100 gammas and the gravity anomaly values range between −60 and −65 milligals (e); with that of the areas in which the aeromagnetic anomaly values range between 2100 and 2500 gammas, and the gravity anomaly values are < −65 milligals (f); with that of the areas in which the aeromagnetic anomaly values range between 2100 and 2500 gammas and the gravity anomaly values range between −60 and −70 milligals (g); with that in which the aeromagnetic anomaly values range between 2100 and 3400 gammas, and the gravity anomaly values are < −70 milligals (h); with that in which the aeromagnetic anomaly values range between 2500 and 3400 gammas, and the gravity anomaly values range between −65 and −70 milligals (i).

109

8.5G. Terrains for which the aeromagnetic anomaly values range between 2100 and 2500 gammas, and the gravity anomaly values range between -60 and -70 milligals: four occurrences.

8.5H. Terrains for which the aeromagnetic anomaly values range between 2100 and 3400 gammas, and the gravity anomaly values are < -70 milligals: seven occurrences.

8.5I. Terrains for which the aeromagnetic anomaly values range between 2500 and 3400 gammas, and the gravity anomaly values range between -65 and -70 milligals: one occurrence.

The results of this application, as shown in the figure, suggest the existence of several environments: different types of base metal mineralization are represented in the data analyzed. The derived patterns provide the opportunity for mentioning some of the critical variables that must be considered in modeling map patterns. The size, shape, and orientation of the occurrence neighborhoods to be used for establishing the initial coincidence with map patterns of geological units are of importance. One might want to experiment with circular, square, rectangular, or elliptical neighborhoods, or with neighborhood sizes smaller than the average size of the outcrops, subparallel to geological trends, or of arbitrary geometry and orientation—which would provide different kinds of controlling relationships in better correspondence with the geological units hosting the mineralization.

In selecting the contour intervals to be digitized from the geophysical maps, several intervals have been lumped together. This may cause the extracted binary patterns to represent overly broad ranges of values. The transformation of continuous variables (like the contour values) into discrete variables (such as the binary patterns extracted here) may result in a loss of information. Only the prior knowledge of the geologist and the geophysicist can guide decisions of this nature.

Additionally, the knowledge of mineralization characteristics of the occurrences used in constructing the controlling neighborhoods and the availability of additional information are essential tools to the model-building experiments.

The patterns derived in this section represent an initial stage of an application that elucidates what sorts of geological decisions often have to be made, even during an interactive session.

CONCLUDING REMARKS

The target of the analysis of binary images of the type considered in this chapter is to recognize patterns of areas in which there is a greater probability of discovering mineral deposits. In the effort to hit the target, we must always keep in mind that in building the necessary geological models for recognition, it is important to decide what to measure from our images, given the various uncertainties associated with the data: (*a*) uncertainties in the geological boundaries represented on the geological maps and in the contours of the geophysical maps; (b) uncertainties in the resolution of the original data mapped or contoured (not all available information can be represented on a map); and (c) uncertainties in the digital resolution of the images, in their registration, and in the selection of the information digitized. These uncertainties are secondary to the greater uncertainty, involved in constructing the classification model mostly on the basis of prior knowledge.

The transformations of the contour intervals in geophysical (or geochemical) maps into binary images and, most of all, the selection of the intervals, have to be performed with expert knowledge of what the geophysical anomalies represent in terms of the phenomena. The contouring itself is, to some extent, an artistic endeavor. According to the different applications, geophysicists produce special-purpose contour maps for their work; for this reason a published geophysical anomaly (or geochemical) contour map is not necessarily the most appropriate database for a given study.

Additionally, it may be remarked that the binary transformation of continuous data represents a gross simplification. It would be desirable to treat contour maps the same as gray level images, therefore deciding on an acceptable method of interpolation. It naturally follows that the applications described here can be only of a general nature: a few fundamental aspects are considered in order to suggest a broader family of promising techniques unfamiliar to geologists.

The results of the experiments described in this chapter have shown that it is possible for a geologist to perform by himself the rapid processing of data on mineral resources from digitized maps. It also has been demonstrated that it is useful to have devices that make probability concepts visible and therefore more understandable by geologists, such as map overlays. It is important that geological maps and ancillary data on resources be transformed and processed as they are in remote sensing. This represents one more step toward the integration of geoscience data. The approach leads to "computerized vision for the geologist" (Kasvand, 1983), in which a computer is used at various levels as an analytical tool and a consultant.

REFERENCES

Bonham-Carter, G. F., and C. F Chung, 1983, Integration of Mineral Resource Data for Kasmere Lake Area, Northwest Manitoba, with Emphasis on Uranium, *J. Math. Geol.* **15:**15-45.

Chung, C. F., 1979, A System of Interactive Graphic Programs for Multivariate Statistical Analysis for Geological Data, in *Proc. 12th Symposium on Interface of Computer Sciences and Statistics,* University of Waterloo, pp. 452-456.

Chung, C. F., 1983, SIMSAG: Integrated Computer System for Use in Evaluation of Mineral and Energy Resources, *J. Math. Geol.* **15:**47-58.

Fabbri, A. G., 1981, Image Processing of Coincident Binary Patterns from Geological and Geophysical Maps of Mineralized Areas, in *Proc. Canadian Man-Computer Comm. Soc., 7th Conf.,* June 10-12, 1981, *Waterloo, Ontario,* pp. 323-331; also in *Uranium in Granites,* Y. T. Maurice, ed., 1982, *Geol. Surv. Can.,* Paper 81-23, pp. 157-165.

Fabbri, A. G., and T. Kasvand, 1981, Applications at the Interface between Pattern Recognition and Geology, *in Sciences de la Terre*, Série Informatique Géologique, no. 15, pp. 87-111.

Kasvand, T., 1983, Computerized Vision for the Geologist, *J. Math. Geol.* **15**:3-23.

Weber, W., D. C. P. Schledewitz, C. F. Lamb, and K. A. Thomas, 1975, *Geology of the Kasmere Lake-Whiskey Jack Lake (North Half) Area (Kasmere Project)*, Manitoba Department of Mines, Resources and Environmental Management, Mineral Resources Division, Geological Sevices Branch, Pub. 74-2, 163p.

CHAPTER 9

Digitization and Preprocessing of Microscopic Images of Rocks in Thin Section

TEXTURES AND SCANNING

Under the microscope, rocks in thin sections or polished sections appear as mosaics of tightly interlocked grain profiles. The sections correspond to projections of the crystalline boundaries onto a plane: the geometrical characteristics of these profiles, including their apparent shapes, orientations, and distributions, belong to what is generally termed texture. The *Dictionary of Geological Terms* (American Geological Institute, 1957), defines *texture* as "geometrical aspects of the component particles of a rock, including size, shape, and arrangement."

A more precise definition of texture and a formal approach to its study do not exist. The characterization of textures represents a complex and unresolved problem; a great variety of different textural aspects and methods for measuring textural properties are being applied in diverse fields of study. In a recent review of approaches to textures in image processing, Haralick (1979) emphasizes that in order to characterize textures, both the tonal primitives — a particular range of gray levels or a single phase, for example — and their spatial interrelationships must be characterized. He considers the following eight groups of statistical approaches: (a) autocorrelation function, (b) optical transforms, (c) digital transforms, (d) textural edgeness, (e) structural element, (f) gray-tone co-occurrence, (g) run lengths, and (h) autoregressive models. As Haralick observes, each of the methods existing to date tends to emphasize either the analysis of tonal primitives or their spatial interrelationships, and does not treat each aspect equally. In describing the structural-element approach — proposed by Matheron (1967, 1975) and by Serra (1978) —

he writes, "The power of the structural element approach is that it emphasizes the shape aspect of the tonal primitives. Its weakness is that it can only do so for binary images" (Haralick, 1979, 787).

Contrary to that impression, the structuring-element approach can be extended to nonbinary images, as exemplified by Serra (1976) and by Goetcherian (1980). This was done by Sternberg (1978, 1980) and Gillies (1978) for the Cytocomputer, a pipeline processor of recent design (Preston et al., 1979). This approach is the target of the digitization and preprocessing techniques described here for microscopic images of rocks.

A microscopic image can be considered a raw gray-level image. From it we can extract a sample (or better, a discrete array of gray-level values) at regular intervals (digitized image). To obtain a colored image, we can scan the image with different color filters. A scanner, either mechanical or electronic, measures the gray-level tones from the original image material and transforms them into digital values. These can be stored for later use, on a magnetic tape, for example. In this instance, for transferring the information provided by the different colors many images will have to be combined. Some crystals, for example, appear colored under plane polarized light whereas other crystals are colorless. The color of pleochroic crystals, however, will change with their orientation with respect to the plane of polarization. Under cross-polarized light, the colorless crystals will assume different colors according to their orientation relative to the polarization planes and also as a result of their individual lattice and optical properties. Under these general circumstance, the recognition of grain identity and grain boundaries becomes a problem even in the relatively simple instances of clear and homogeneous crystalline fabrics.

For example, whenever crystals of the same type (phase) are in contact, the grain boundary may not be visible until the microscope setting is changed. Even when an already available colored draft is being scanned, the scanner cannot resolve the colors easily as uniform gray tones because colors may be made up of colored dots, and there are defects even in colors that look uniform. For these reasons it is not yet possible for automatic scanning devices to capture and process sufficient information for satisfactory phase recognition and extraction.

Several types of scanners can be used to digitize thin-section material: the flying spot Scanner (FSS), the television scanner (TV camera), the drum scanner, the flatbed scanner, and the scanning microscope (microdensitometer).

The FSS is a custom-made electronic device with normal resolution ranging from 1024 pixels × 1024 pixels to 4096 pixels × 4096 pixels. Its position accuracy is not good because it uses a magnetically deflected electron beam. The scanning speed is equivalent to computer speed if the FSS is controlled by a computer.

The broadcast commercial television camera (scanner) is an electronic device with resolution of 525 lines in North America and 625 lines in Europe (for the PAL system). The aspect ratio is 4/3. Special TV cameras have up to 1200 or more lines of resolution. This scanner is the least expensive and also the fastest: it grabs the input in 1/30 second. Although there are problems of position accuracy with electronic television scanners, there are not with solid-state chips television

114

cameras.Unfortunately, however, 500 × 500-pixel "chips"are not yet available at reasonable cost in solid-state TV cameras.

Mechanical drum and flatbed scanners are not limited in resolution, except for the physical limits of the particular equipment. They are position-accurate, but they are slow: for example, what could take ten minutes on a drum scanner might require two hours on a flatbed scanner. They are expensive devices, particularly the flatbed scanners.

A scanning microscope consists of a television camera, which looks at the image in the microscope, and a single photocell (a photomultiplier for photometric accuracy): these complement one another. The television camera has a low resolution, position accuracy, and a fast speed. The photocell has a high resolution and good position accuracy, but it is slow. Scanning microscopes are also expensive instruments.

At present, the optimal and easiest method of image processing, and even the quickest, is for the operator to trace the boundaries of crystal profiles. The application considered in this chapter makes use of a flying spot scanner, although the grain-profile boundaries of a microscopic image could also be digitized by projecting the microscopic image onto a graphic tablet. The latter method, however, would have required some special equipment (a transparent tablet digitizer and projection devices) and also would have been too time-consuming on a single-user computer (several thousand grain profiles would have to be recognized and manually digitized).

This chapter describes in detail the digitization of ink drawings of grain profiles by a flying spot scanner. Chapters 10 and 11 will discuss experimental applications of the analysis of the images of a granulite and an amphibolite, and thus will exemplify the capabilities of the software which was programmed in GIAPP. They also provide the ground for further work on the quantitative characterization of rock textures.

DIGITIZATION OF GRAIN PROFILES BY SCANNING TRANSPARENCIES OF LINE DRAWINGS

The procedures described in this section could be automated to some extent, by using a more expensive scanner than the one used. This description faithfully represents the experiments and the processing actually performed for the practical applications.

In summary, the steps for digitizing the microscopic images of grain profiles from the thin sections are as follows:

1. The original microscopic image is traced by projecting it onto a wall or a special projecting screen. A 35-mm negative drawing of such a tracing is shown in Figure 9.1A.

2. The image is scanned first within a properly positioned box (area selected for scanning), and displayed on a Tektronix 611 storage display screen, as seen in

115

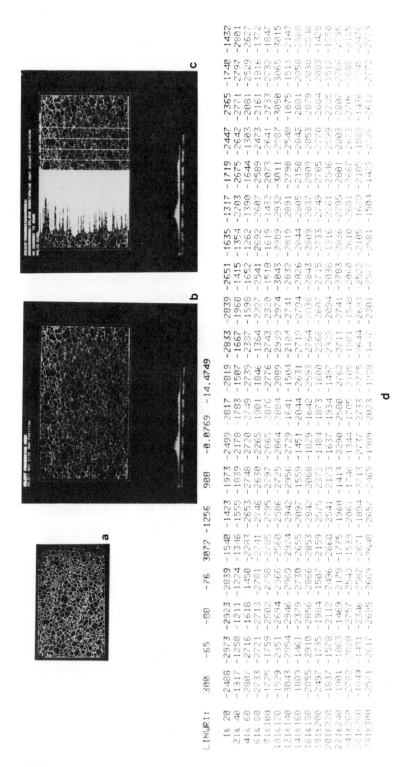

Figure 9.1. (a) Negative print of the line drawing of 1300 grain profiles redrafted after Kretz (1969); (b) Tektronix 611 video display plot of the scanned image, and of the gray levels; (c) same plot of part b with the cross-section plot of the gray levels; (d) printout of the gray-level readings displayed in the cross-section in part c.

Figure 9.1B. A histogram of the gray-level value readings is produced, also shown in Figure 9.1B. A cross-section of the gray levels between two selected points can be obtained and plotted over the scanned image plot, for further aid in deciding on a proper resolution, as shown in Figure 9.1C. A printout of 256 gray-level readings along the cross-section can be produced; see, for instance, Figure 9.1D. From these preliminary tests, the desired resolution can be determined: for example, at least three pixels should represent line thickness and most small areas should be represented by at least one pixel.

3. After the resolution is determined, the FSS is driven by the computer to scan the transparency and to provide the computer with the desired number of gray-level readings for the desired number of scan lines. In the example shown in Figure 9.2, a 1000 × 595-pixel square raster image was produced: the gray-level values of a few pixels are displayed between x-coordinates 9 and 33, and between y-coordinates (rows) 1 and 45. A second hexagonally scanned image for the same resolution (same density of scanned points per unit distance in the horizontal direction) was also produced, with 1000 pixels × 687 pixels. Each pixel corresponds to 1/53 (0.019) mm.

4. The gray-level image is analyzed by producing first a gray-level histogram, both as plot and as listing, as shown in Figures 9.3A and 9.3B. This histogram is needed to decide how best to select a thresholding level or thresholding rule, also termed gray-level slicing, for obtaining a corrected binary image. In such an image, all the pixels belonging to the lines are black (gray-level value of 1), and all the remaining pixels are white (grey-level value of 0). For this, one or more cross-sections are computed across the image and the gray-level values along them are plotted to see if the desired detail is expressed properly. (Fig. 9.4B). In this case the cross-section has a concave pattern, which means that the scanner sensitivity varies from the center to the edges of the scanning area.

5. Different thresholding levels can be set for producing binary images and studying the plots of relevant details in critical parts of the binary images obtained. Square and pseudohexagonal plots of binary images thresholded at the same values can be produced for comparison of the detail digitized by the two types of raster with identical resolution. These are shown in Figure 9.4 for both the upper left and the upper right corners.

6. To correct for the fact that, in the particular scanner used, sensitivity varied from the center to the edges of the scanning area, high-pass filtering was used. (See Andrews (1970) for a description of the use of filters for image enhancement). In this instance the equivalent effect of high-pass filtering is obtained by computing first the average values for 19 × 19-pixel areas and then creating an image in which the values of the pixels are the differences between the original values and the average values. Further filtering was

117

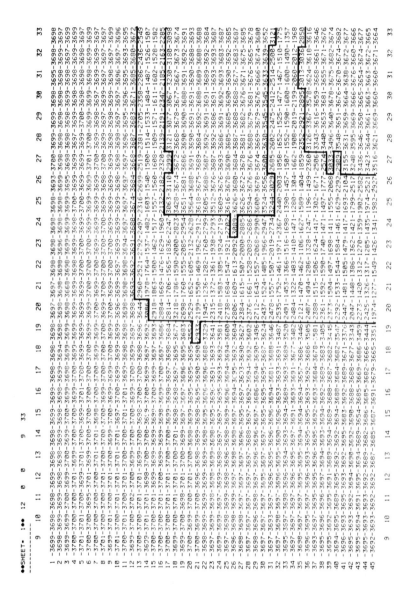

Figure 9.2. Portion of the square raster-scanned image of 1300 grain-boundary profiles of the same granulite shown in Figure 9.1a. The dimensions of the original image are 1000×595 pixels; a 23×45-pixel portion is displayed here as a printout of the gray-level readings. Contours were added by hand to emphasize the location of the boundary pixels.

118

TRAJD2: LHIS(X) =   X=1- 256   X & Y GRID SPACE= 100 100   Y SCALE= 0.93903E-01   NONSCALED MAX.MIN= 0.10100E+05   0.00000E+00

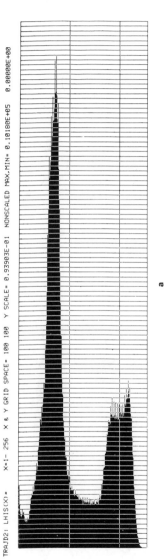

a

b

RAUH3H   17   1000   595   256   -3702   596   -3702

0.000612

Range	from	to											
1 TO 10	-3702	11.) -3589	1719	1305	891	748	663	637	668	798	689	729
11 TO 20	-3589	11.) -3476	755	657	567	537	613	555	529	630	567	601
21 TO 30	-3476	11.) -3363	619	758	759	884	1026	1050	1125	1187	1268	1153
31 TO 40	-3363	11.) -3250	1168	1348	1379	1404	1543	1750	1610	1701	2016	2087
41 TO 50	-3250	11.) -3137	2236	2375	2665	2638	2538	2917	2780	2940	3079	3415
51 TO 60	-3137	11.) -3024	3298	3396	4078	4207	4616	5298	6466	6359	7007	8096
61 TO 70	-3024	11.) -2912	7633	7808	7986	8927	8317	8513	9332	8818	8844	9076
71 TO 80	-2912	11.) -2799	10082	9387	9426	10180	9352	8867	8799	9187	7837	7200
81 TO 90	-2799	11.) -2686	6998	5679	5056	4382	4374	3511	3181	3149	2621	2461
91 TO 100	-2686	11.) -2573	2235	2298	1911	1848	1871	1599	1487	1473	1577	1395
101 TO 110	-2573	11.) -2460	1347	1467	1238	1215	1198	1273	1195	1180	1190	1104
111 TO 120	-2460	11.) -2347	1056	1118	1203	1084	1074	1093	1046	1051	1006	1113
121 TO 130	-2347	11.) -2234	996	985	1038	986	975	970	1053	899	973	1050
131 TO 140	-2234	11.) -2122	957	916	944	1037	935	928	1043	944	915	952
141 TO 150	-2122	11.) -2009	1012	917	875	986	919	884	912	965	955	917
151 TO 160	-2009	11.) -1896	932	902	888	893	954	953	927	1013	902	960
161 TO 170	-1896	11.) -1783	1028	1106	1081	1064	1209	1204	1322	1405	1622	1651
171 TO 180	-1783	11.) -1670	1654	1928	1943	1388	1289	1346	2312	2440	2641	2583
181 TO 190	-1670	11.) -1557	2616	2570	2925	2735	2140	2756	2712	2709	2706	3016
191 TO 200	-1557	11.) -1444	2731	2651	2989	2730	2756	3100	2685	2641	2654	2959
201 TO 210	-1444	11.) -1332	2662	2676	2939	2939	2662	2677	3001	2801	2720	2804
211 TO 220	-1332	11.) -1219	3159	2997	3306	3306	3061	3153	2964	3100	3231	3105
221 TO 230	-1219	11.) -1106	3313	2965	2765	2653	2723	2262	1929	1861	1473	1248
231 TO 240	-1106	11.) -993	1032	973	760	590	476	360	250	250	164	117
241 TO 250	-993	11.) -880	67	46	29	27	14	9	8	8	5	2
251 TO 256	-880	11.) -813	0	3	0	0	0	1				

Figure 9.3. (a) Gray-level histogram of the scanned image in Figure 9.2; (b) list of the gray-level values in the histogram in part a.

119

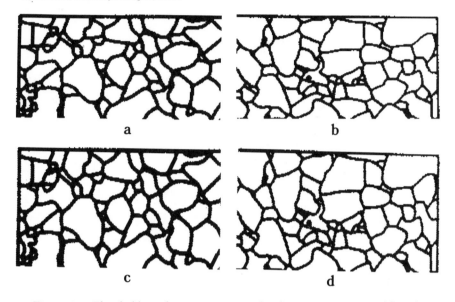

Figure 9.4. Thresholding of two images scanned with a square raster, *a* and *b*, and with a hexagonal raster, *c* and *d*. Thresholding is below the value of −1900 for both the images. Arrays of 2 × 2 black dots correspond to the black pixels. The pseudohexagonal plots in *c* and *d* are obtained by shifting one dot to the right for every even row of image data.

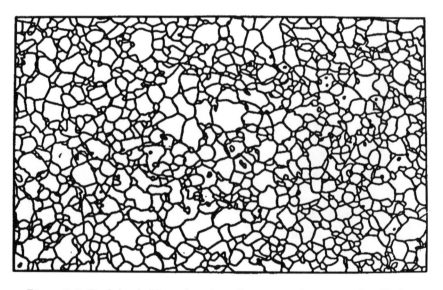

Figure 9.5. Final thresholding selected at values greater than or equal to 65; the entire binary image is displayed.

obtained by reaveraging the averaged values, for example, averaging within 19×19-pixel arrays again. A better filtered image was obtained by subtracting the latter averaged values from the original gray-level values. This second filtering eliminates entirely the concavity of the gray-level value distribution observed in cross-sections of the image (not shown here).

7. After high-pass filtering, several thresholding values are tested again, and are displayed as plots of parts or whole of the binary images obtained. A satisfactory thresholding interval is then selected, as shown in Figure 9.5 or in Figure 9.6A.

8. The binary image of grain-boundary profiles is then edited interactively in places of poor resolution: for example, black pixels are "added" where boundaries are broken, and white pixels are "added" to areas totally filled with black pixels. This interactive procedure, described in more detail in Chapter 6, produces an edited binary image (Fig. 9.6B).

9. The edited, thresholded, and filtered image is thinned so that boundary lines become one pixel wide. The process produces images of the kind displayed in Figures 9.6C and 9.6D.

10. Minor editing might be required in places omitted during the editing step (8) and where the thinning process leaves a few tails. A final boundary image is then obtained (Fig. 9.7).

11. Further processing requires that a unique sequential label be assigned to each 0 pixel entirely surrounded by 1 boundary pixels. This process is termed *component labeling*. The image is not any more binary, as shown in Figures 9.8A and 9.8B.

12. An interactive procedure, identical to the one described in Chapter 6, assigns to all pixels belonging to each different area, different grain, a new phase label, or crystal-type label. A portion of such a phase-labeled image is shown in Figure 9.8C.

13. A phase-labeled image is therefore produced on magnetic tape, which can be transferred to another computer for further computations.

14. The phase-labeled image represents a data bank from which many measurements can be made and many features extracted. For example, a binary image of each phase, or crystal type, can be extracted for further processing. Eight binary images have been obtained (Fig. 9.9) which correspond to the eight phases labeled in the image.

121

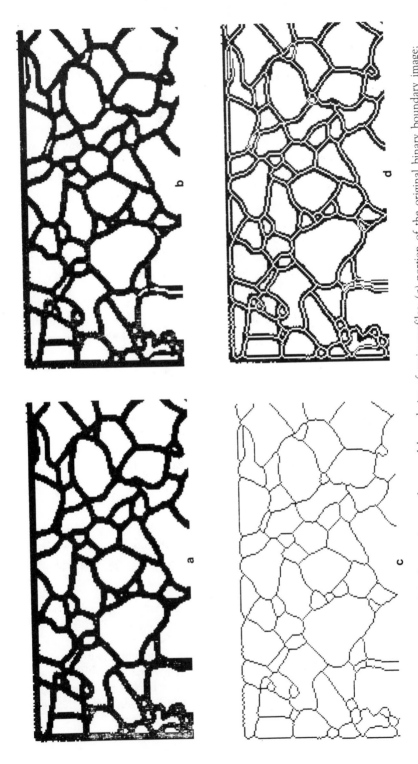

Figure 9.6. Editing of the binary image of scanned boundaries of grain profiles: (*a*) portion of the original binary boundary image; (*b*) portion of the binary image after a first editing stage; (*c*) line thinning of edited binary image of boundaries; (*d*) black pixels that have been changed to white pixels during line thinning.

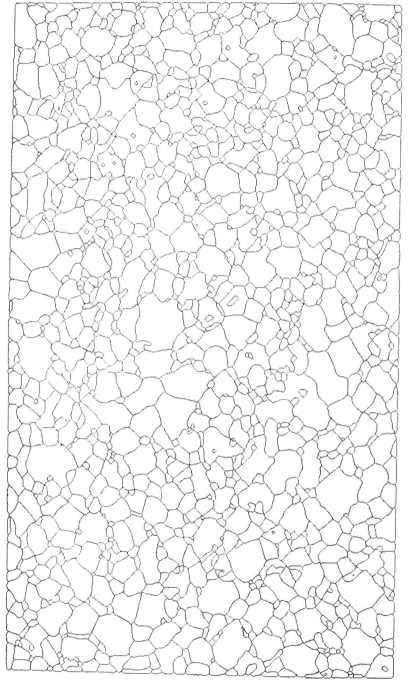

Figure 9.7. Final edited and thinned binary image of boundaries of grain profiles of the granulite.

With the completion of step 14, we have terminated the task of digitization and preprocessing, which is a prerequisite to the analysis of binary images to be described in Chapters 10 and 11. Clearly, steps 3 to 10 could be automated to a great extent by using a more expensive flying spot scanner. If that were the case, the steps would be required only for minor improvements.

From the analysis of binary images of grain profiles, the study of properties in the third dimension is feasible, which is a typical stereological problem. Texture characterization, as performed in Chapters 10 and 11, is only a small aspect of the broader field of texture analysis.

Some new problems can be studied if we assume that we have a computer image of our study material at our disposal, which can be identified as a data bank from which we can read or retrieve information on many quantitative aspects. Indeed, in such a situation— in the absence of a comprehensive theory of texture formation and characterization— what we can ask the data bank is limited only by practical knowledge and ingenuity.

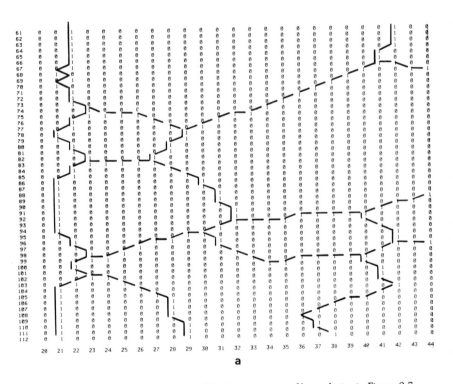

Figure 9.8. Further preprocessing of the binary image of boundaries in Figure 9.7: (*a*) portion of expanded binary image of boundaries; (*b*) component-labeled image; (*c*) phase-labeled image.

124

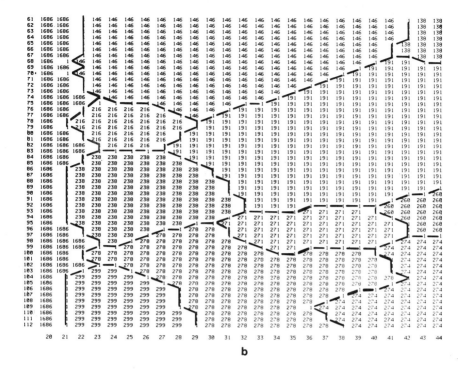

b

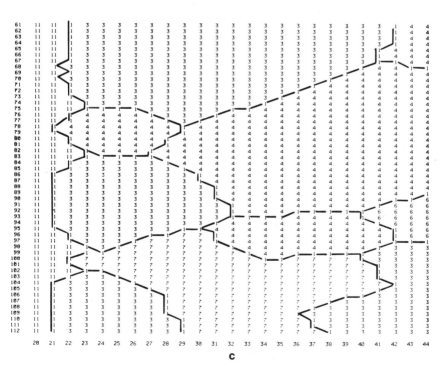

c

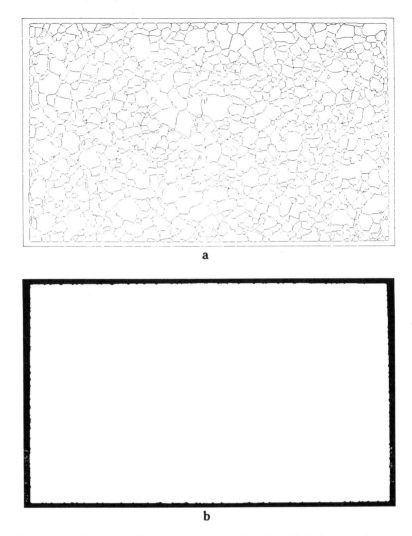

a

b

Figure 9.9. Extraction of binary images from the phase-labeled image of grain-boundary profiles of granulite. The image dimension is 1000 × 595 pixels. (*a*) image of edited thinned boundaries, with 38,225 black pixels; (*b*) image of the frame, 87,100 black pixels; (*c*) image of the pyroxene profiles, 278,010 black pixels; (*d*) image of the scapolite profiles, 204,741 black pixels; (*e*) image of the sphene profiles, 10,909 black pixels; (*f*) image of the hornblende profiles, 12,360 black pixels; (*g*) image of the apatite profiles, 1,835 black pixels; (*h*) image of the zircon profiles, 45 black pixels.

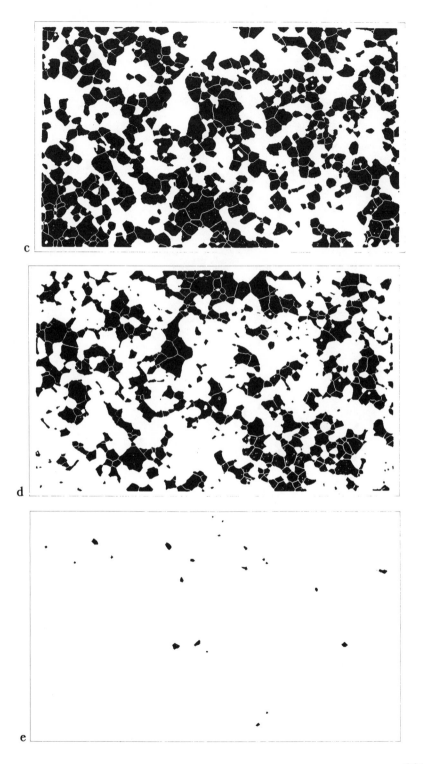

c

d

e

127

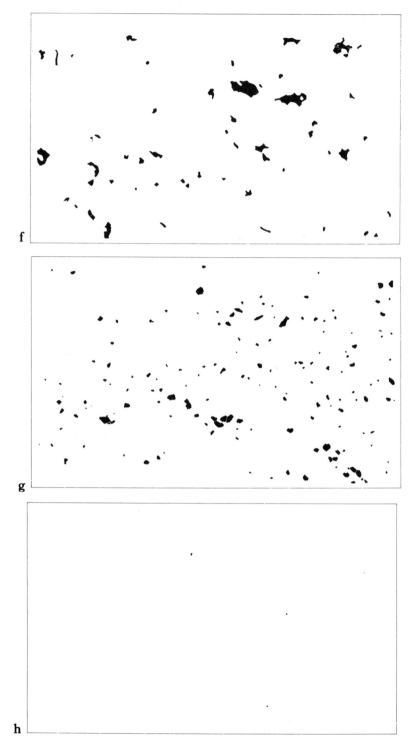

f

g

h

128

REFERENCES

American Geological Institute, 1957, *Dictionary of Geological Terms*, Doubleday, Garden City, N. Y., 575p.

Andrews, H. C., 1970, *Computer Techniques in Image Processing*, Academic, New York, 187p.

Gillies, A. W., 1978, An Image Processing Computer Which Learns by Example, in *Image Understanding Systems and Industrial Applications*, R. Nevatia, ed., Proc. of the Society of Photo-Optical Instrumentation Engineers (SPPIE), Aug. 30-31, 1978, San Diego, Cal., vol. 155, pp. 120-126.

Goetcherian, V., 1980, From Binary to Gray Tone Image Processing Using Fuzzy Logic Concepts, *Pattern Recognition* **12:** 7-15.

Haralick, R. M., 1979, Statistical and Structural Approaches to Texture, *IEEE Proc.* **67:** 786-804.

Kretz, R., 1969, On the Spatial Distribution of Crystals in Rocks, *Lithos* **2:** 39-65.

Matheron, G., 1967, *Eléments pour une théorie des milieux poreux*, Masson et Cie, Paris, 166p.

Matheron, G., 1975, *Random Sets and Integral Geometry*, Wiley, New York, 261p.

Preston, K., Jr., M. J. B. Duff, S. Levialdi, P. E. Norgren, and J-i. Toriwaki, 1979, Basics of Cellular Logic with Some Applications in Medical Image Processing, *IEEE Proc.* **67:** 826-858.

Serra, J., 1976, *Lectures on Image Analysis by Mathematical Morphology*, Cahier N-475, Centre de Morphologie Mathematique, Fountainebleau, France, July 1976, 225p.

Serra, J., 1978, One, Two, Three, Infinity, in *Geometrical Probability and Biological Structure: Buffon's 200th Anniversary*, R. L. Miles and J. Serra, eds., Springer-Verlag, New York, pp. 137-152.

Sternberg, S. R., 1978, Cytocomputer Real-Time Pattern Recognition (abst.)., in *8th Annual Automatic Imagery Pattern Recognition Symposium Proceedings on "Emerging Pattern in AIPR,"* National Bureau of Standards, Gaithersburg, Maryland, p. 205.

Sternberg, S. R., 1980, Language and Architecture for Parallel Image Processing, in E. S. Gelsema and L. N. Kanal, eds., *Pattern Recognition in Practice*, Amsterdam, North-Holland Publishing Company, p. 35-44.

10

Some Aspects of the Quantitative Characterization of a Thin Section of a Granulite

As previously mentioned, the tracing of the profiles of approximately 1300 grains of crystals in a thin section of a granulite was obtained from Kretz (1969). He was the first to give particular attention to the statistical analysis of grain profiles for quantitatively relating their geometrical attributes to nucleation and crystallization processes that could be modeled for this rock. Kretz used several manual methods for obtaining the measurements from a complete drawing of a thin section (Kretz, 1969, Fig. 1). The same measurements, and many more, can be made from the drawing, once it is in digital form, that is, once it is computer processible, as is the explicit image shown in Figures 9.8C and 9.9.

Only some experiments are described here, particularly those that would be too cumbersome to perform manually. They are included to exemplify the kind of studies that are possible with relative ease by programming and processing images on a small computer.

The applications described in this chapter involve (a) computation of the area and the circumference (perimeter) of the grain profiles; (b) measurement of grain-profile contacts and of their distribution; (c) determination of the orientation of grain and grain-cluster profiles; and (d) computation of the geometrical covariance function of the fabric. These applications explain how to study a rock fabric systematically, but they also provide some new unconventional geological tools for detecting the presence of, and describing the type of, crystal shape or crystal cluster anisotropy that causes gneissosity or foliation in a crystalline fabric.

COMPUTATION OF THE AREA AND CIRCUMFERENCE OF GRAIN PROFILES

In general, the basic needed geometrical data from a thin section are the area and the perimeter length (circumference) of all the grains belonging to each phase. A summary of data for the Grenville granulite is given in Table 10.1. There, because of the particular partitioning of the image data, separate areas are given for the boundary (see Fig. 9.9A) and the frame (see Fig. 9.9B) and also for the boundary pixels in contact with the frame. Because the boundary itself is distinguished as a separate phase, one method of calculating the areal proportion, in percent of the crystalline phases, is to divide the number of pixels belonging to each phase by the sum of all the pixels belonging to crystalline phases: 507,900 pixels. The six percentages so computed in column 2 of the table can be compared with the percentages, in column 3, obtained by Kretz (1969) for a portion of the draft used here for digitizing the image of the granulite. The numbers of crystals, the crystal proportions, and the average area per crystal are also given in the table (in columns 4 and 1, respectively). The latter data show that the pyroxene profiles have the largest grain size, followed by scapolite, amphibole, apatite, sphene, and zircon.

The total circumference of the grain profiles was computed as follows:

1. From the image of boundaries in Figure 10.1A, the pixels at the contact with the frame (shown in Figure 9.9B) have been "eliminated" by dilatating the frame image with a 3×3 black structuring element, and then computing the image in which the boundary pixels do not overlap with the dilatated frame image, as shown in Figure 10.1B.

2. The separate overlaps between the dilatated phases and the boundary image in Figure 10.1B have produced the images of grain-profile boundaries shown in Figures 10.1C to 10.1H. The number of pixels belonging to each boundary is listed in column 5 of Table 10.1.

3. Such boundary images, however, do not take into account the fact that the boundaries between grain profiles of the same phase have to be measured once for each of the adjacent grain profiles. The separate contacts between adjacent grain profiles of the same phase can be obtained, for example, by "closing" each phase and by computing the intersection or overlap between the closed image and the boundary image.Figure 10.2A shows a pyroxene-pyroxene boundary and Figure 10.2G a scapolite-scapolite boundary. The number of pixels on the boundaries between the profiles of the same phase are also given, in brackets, in column 5 of Table 10.1.

4. The boundary lengths were computed by "cross-correlating" each partial boundary image with the total boundary image in Figure 10.1B, for shifts in vertical and horizontal directions of one pixel in both senses. The values for shifts $(-1, -1), (+1, -1), (+1, +1)$, and $(-1, +1)$ were summed, divided by 2,

132

TABLE 10.1
Summary of Quantitative Characterization Data of the Grain Profiles
in the Granulite

	1	2	3	4	5	6	7†	
Phases	No. of pixels (Average No. per Crystal)	Area (%) over 507,900 Pixels	Area (%) from Kretz (1969)*	No. of crystals (Proportion)	No. of Pixels on boundaries (Between Crystals of Same Phase)	Circumference	Complexity Index (C.I.) a (b)	c (d)
Frame	45,828							
Frame/boundary contact	3,047							
Boundary	38,225							
Pyroxene	278,010 (592)	54.74%	50.31%	470 (0.363)	28,579 (8,136)	47,495	0.15 (0.17)	0.14 (0.22)
Scapolite	204,741 (357)	40.31	41.89	574 (0.444)	26,627 (7,123)	44,041	0.18 (0.17)	0.17 (0.19)
Sphene	10,909 (67)	2.15	3.38	164 (0.127)	4,296 (99)	6,012	0.39 (0.54)	0.37 (0.43)
Amphibole	12,360 (203)	2.43	4.05	61 (0.047)	3,107 (114)	4,202	0.27 (0.33)	0.22 (0.25)
Apatite	1,835 (83)	0.36	0.34	22 (0.017)	668 (−)	912	0.36 (0.50)	0.33 (0.39)
Zircon	45 (15)	0.01	0.03	3 (0.002)	44 (−)	64	0.72 (1.4)	0.50 (0.92)
Crystal phases only	507,900	100.00%	100.00%	1,294 (1.000)	63,321 (15,472)	102,726		
Total image	595,000							

a: C.I. with boundary pixels as part of crystals
b: C.I. without boundary pixels as part of crystals
c: C.I. for circle of same average area computed as in (a)
d: C.I. for circle of same average area computed as in (b)
* For a portion of the study area.
† Note on C.I. computations: for pyroxene, for example, (the complexity index computations in column 7 are as follows: (a) 47,495/(28,579 + 8,136 + 278,010) = 0.15. (b) 47,495/278,010 = 0.17. (c) For A of (28,579 + 8136 + 278,010)/470 = 670, $r = 14.60$, $c = 91.76$, and $c/A = 0.1370$, or 0.14 (d) For A (area of circle) = πr^2, and c (circumference) = $2\pi r$, $c/A = 2/r =$ C.I. For A of 278, 010/470 = 592, $r = 13.73$, c-129.38, and $c/A = 129.38/592 = 0.2185$, or 0.22.

and the result multiplied by 2. This value, the total length of the boundary in 45° and 135° directions, was added to the total length of boundary in 90° and 180° directions. This was computed as the sum of the values for shifts $(0, -1)$, $(+1,0)$, $(0,+1)$, and $(-1,0)$ divided by $\sqrt{2}$. (This procedure was also used in Chapter 4.) The sums of the lengths of the boundary between grain profiles of all phases and those of grains of the same phase is the final length entered in Table 10.1, column 6.

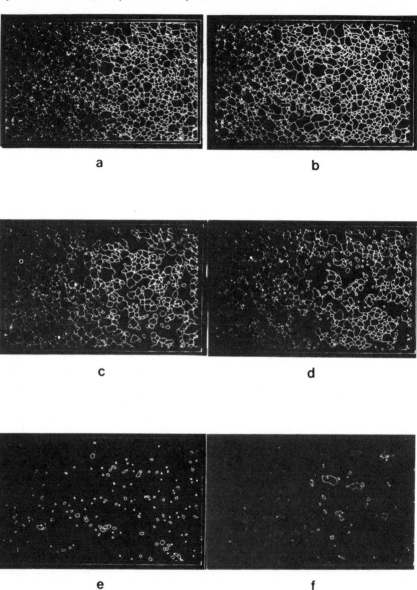

Figure 10.1. Partitioning of the grain profile boundary image of the granulite into images of perimeters of the individual phases: *(a)* image of phase boundaries with frame/boundary contact; *(b)* image of phase boundaries without frame/boundary contact; *(c)-(h)* images of the individual boundaries: pyroxene*(c)*, scapolite*(d)*, sphene*(e)*, amphibole (hornblende)*(f)*, apatite*(g)*, and zircon*(h)*. The numbers of boundary pixels in these images and the boundary lengths are listed in Table 10.1.

g   h

The "complexity index" (C. I.) is a geometrical measure that relates the circum-ferences and the areas of the grain profiles for eash phase (Underwood, 1970, 228-229). It should reflect the jaggedness of the perimeter per area of the particles or grain profiles. This is simply computed as the ratio between the total circumfer-ence and the total area of the grain profiles for each phase (see column 7 in Table 10.1). Because of the particular nature of the data in this application—binary images in which the boundary is a separate phase—this index can be affected by the resolution so that the phases with the smaller grains tend to have a relatively larger circumference. This is the case for zircon (column 7 of Table 10.1). For this reason, two different computations for the C. I. have been entered in the table: one in which the boundary pixels have been summed to the areas of the grain profiles, and one in which only the areas of the grain profiles were used. In the table we can see that in general the larger grains have smaller C. I.'s. In column 7 the C. I.'s computed for circles of the same average areas (number of pixels per crystal, listed in column 1) are also shown. According to Underwood (1970), the C. I. can be computed without assumption of grain size.

MEASUREMENT OF GRAIN-PROFILE CONTACTS AND THEIR DISTRIBUTION

Binary images of the separate grain-profile contacts can be obtained for each phase, by computing the intersections between all possible pairs of the boundary images shown in Figure 10.1C to 10.1H. Nine of the most revealing partitioned boundary images are shown in Figure 10.2. Intraphase boundaries, Figures 10.2A and 10.2G, are characterized by the presence of junctions between boundary segments, which occur wherever three profiles of grains of the same phase are in contact. The other types of boundary images are typically of "meander"-like appearance, with different degrees of dispersion or clustering in relation to those of the grain profiles. Of particular interest is the boundary between the two main phases pyroxene-scapolite, (Fig. 10.2B), which strongly characterizes the entire fabric.

135

Figure 10.2. Nine binary images of separate grain-profile contacts in the granulite: *(a)* pyroxene-pyroxene, *(b)* pyroxene-scapolite, *(c)* pyroxene-sphene, *(d)* pyroxene-amphibole, *(e)* pyroxene-apatite, *(f)* pyroxene-zircon, *(g)* scapolite-scapolite, *(h)* scapolite-sphene, and *(i)* scapolite-amphibole.

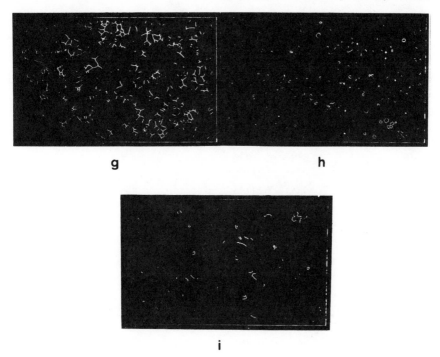

g

h

i

As previously mentioned, the length of each extracted boundary image can be computed, and a table of boundary length distribution is constructed, as shown in Table 10.2. In part A of the table, both the number, n, of pixels on the boundary, and the boundary length, l, are entered for comparison. From both kinds of data, a boundary length distribution for a compressed array of boundaries (contacts) pyroxene (P), scapolite (S), and other phases (O) are easily constructed and compared with the expected values for boundary distributions for the supposition that the crystals are randomly distributed, which are shown in brackets in part B. From these arrays, two different transition matrices are computed, **P(n)** and **P(l)**, respectively in part C, in which the sums of each row equal unity. The transition matrices represent the probability that a grain belonging to a phase is in contact with a grain of the same or of any other different phase. As explained by Kretz (1969, 59, eq. 4) we can obtain the transition matrix based on the supposition of random distribution of crystals by the following equation which is valid if events pyroxene, P, and scapolite, S, are independent:

$$p(P/S) = p(P) \text{ and } p(P/P) = p(P)$$

where $p(P/S)$ is the probability of event pyroxene, given scapolite, and $p(P)$ is the probability of event proxene alone.

137

TABLE 10.2A
Distribution of Boundary Pixels, n and Boundary Lengths, l (in parentheses),
Between All Pairs of Phases of the Granulite

	P	S	T	H	A	Z	
P	8,136 n (10,571)						
S	17,231 (22,525)	7,123 (9,387)					
T	2,281 (3,174)	2,088 (2,944)	99 (136)				
H	2,072 (2,694)	986 (1,351)	125 (185)	114 (156)			
A	300 (423)	365 (510)	34 (49)	20 (28)	— (—)		
Z	25 (38)	20 (34)	— (—)	— (—)	6 (9)	— (—)	
Total	30,045 (34,425)	27,813 (36,751)	4,627 (4,488)	3,317 (4,414)	725 (1,019)	51 (91)	41,025 (54,214)

Note: Letters P through Z indicate the phases pyroxene through zircon.

TABLE 10.2B
Distribution Recomputed for Compressed 3 × 3 Arrays of Boundary
Data, n, and l

n	P	S	O	Sum
P	8,136 (13,610)	17,231 (12,589)	4,678 (3,846)	30,045
S	17,231 (12,599)	7,123 (11,654)	3,559 (3,560)	27,813
O	4,678 (3,866)	3,459 (3,576)	398 (1,092)	8,535
Sum	30,045	27,813	8,535	66,393

l	P	S	O	Sum
P	10,571 (17,781)	22,525 (16,401)	6,329 (5,244)	39,425
S	22,525 (16,575)	9,387 (15,288)	4,839 (4,888)	36,751
O	6,329 (5,291)	4,839 (4,880)	563 (1,560)	11,731
Sum	39,925	36,751	11,731	88,407

Note: The less frequent boundaries have been grouped under O, other phases. Observed values are above and the expected values are below (in parentheses).

TABLE 10.2C
Transition Matrices Computed and Compared with the Transition Matrix
$P(k)$ **Obtained by Kretz (1969)**

(0.271 0.573 0.156)	(0.268 0.571 0.161)	(0.389 0.466 0.145)
$P(n) = (0.620\ 0.256\ 0.124)$	$P(l) = (0.613\ 0.255\ 0.132)$	$P(k) = (0.536\ 0.384\ 0.080)$
(0.548 0.405 0.047)	(0.540 0.412 0.048)	(0.613 0.290 0.097)

$\pi(n) = (0.453\ 0.419\ 0.128)$ $\pi(l) = (0.451\ 0.416\ 0.133)$ $\pi(k) = (0.474\ 0.412\ 0.113)$

Note: The fixed vectors n, l, and k represent the expected transition matrix based on the supposition that the crystals are randomly distributed. They represent the proportions of the sums for n and l in Table 10.2B. The elements of those vectors are multiplied by the sums in Table 10.2B to compute the expected values in terms of boundary pixels and boundary length distributions. See text for additional explanation.

The two matrices $P(n)$ and $P(l)$ are almost identical, which suggests that the computation of boundary length is not required to compute the transition matrices. In Table 10.2C these matrices can be compared with the transition matrix $P(k)$ obtained by Kretz (1969), using the equation above, from a portion of the draft for the same rock. Given the similarity between the three matrices, the same statistical tests and the same conclusions drawn by Kretz, can be made, i.e., that "fewer P-P and S-S transitions and more P-S were found than expected, owing possibly to departure from randomness in the direction of a more regular distribution for pyroxene and scapolite" (1969, 61). A chi-square test to determine if the differences between expected and observed values are significant makes the model of randomness in the distribution of the two crystals just acceptable at the 95-percent level of significance.

DETERMINATION OF THE ORIENTATION OF GRAIN AND GRAIN-CLUSTER PROFILES

A granulite is a high-grade metamorphic rock consisting of even-sized interlocking grains with a weak preferred orientation. The orientation of the fabric can be expressed either as an anisotropic distribution of grains of both major and minor constituents or as a shape anisotropy of the grains and therefore of the grain profiles. In crystalline fabrics all possible interrelationships of these two types of anisoptropies can be observed in nature for the same texture. The latter characteristic is measured in this section.

A method of measuring shape anisotropy consists in computing the rose diagram of the boundaries of grain profiles for each phase. A simple but laborious way to study the rose diagram of the boundaries is to approximate them by successive straight-line segments that are sufficiently short, and to plot the histogram of the combined length of all line segments pointing in directions bounded by class limits a few degrees apart. An alternative method to determine the preferred

139

orientation consists of first obtaining measurements for narrower class intervals, one or two degrees, and to construct a moving average for wider classes to reduce the random fluctuations that generally arise when the classes are too narrow.

A computer program termed RODIA (ROse DIAgram)–developed by Agterberg (1979), and also used by Agterberg and Fabbri (1978) and by Agterberg et al. (1981)– constructs a smoothed histogram for the contact between the two phases ("black" and "white") in a binary image. The input for RODIA is the so-called geometrical covariance of a binary image, which can be computed by GIAPP.

A number of experiments have been performed to produce the rose diagram of the following images: (*a*) the boundaries of the individual grains of pyroxene and of scapolite profiles, (*b*) the boundaries of the clusters of grains of the two phases, (*c*) the boundaries of all the individual grains in the granulite, and (*d*) the "skeletons" of the pyroxene grain profiles.

Some preprocessing was required to produce the binary images input for the computation of the geometrical covariance in two dimensions. For studying the individual grains, the image of octagonally eroded pyroxene grains and scapolite grains, and of all the grains in the granulite (of sizes greater than that of the octagon used for the erosion) were computed. They are shown in Figures 10.3A, 10.3B, *and* 10.3C, respectively. The 3 × 3 square dilatation of the scapolite and pyroxene images was also computed for studying the grain-profile clusters. The image of the dilatated pyroxenes is shown in Figure 10.3D. A rectangular binary mask was produced, with a dimension of 521 × 901 black pixels, at least 5 pixels away from the edges of the frame of the image. The mask is contained in a 1000 × 595-pixel binary image, shown in Figure 10.3E. The intersection of this "mask image" and the images to be analyzed was computed for the covariance measurements. In Figure 10.3F, the image is shown of the intersection between this mask and the image of the octagonally eroded grains in the granulite (i.e., between the images in Fig. 10.3C and those in Fig. 10.3E). The geometrical covariance was computed by shifting the masked images (similar to the one in Fig. 10.3F) for 5 pixels, a pixel at a time, to the right, to the left, and downward, and then computing the intersections between the shifted image and the unshifted image (like the one in Fig. 10.3C). The resulting array of intersections from the image in Figure 10.3F is shown in Figure 10.3G. It consists of 66 values, 61 of which are used for the computations. Similar arrays were obtained for the other experiments previously mentioned.

The rose diagrams for the experiments performed are shown in Figure 10.4, in the form of histograms of directions. The frequency is plotted vertically. The corresponding direction is plotted for clockwise rotation in the horizontal direction starting from 0 for the horizontal left-right direction. Consequently, 90 is for the top-to-bottom vertical direction, and 180 is for the horizontal right-left direction. All the orientation patterns in parts A to G of Figure 10.4 show two maxima for 45° and 130° directions. The pattern for square dilatated pyroxenes (PYSD1) is

140

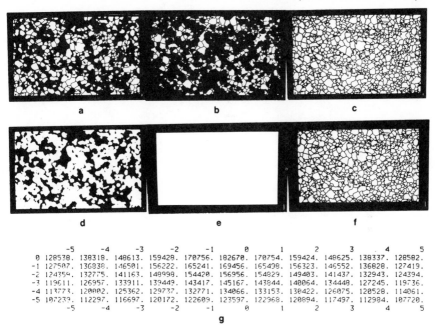

```
         -5      -4      -3      -2      -1       0       1       2       3       4       5
 0  128538. 138318. 148613. 159428. 170756. 182670. 170754. 159424. 148625. 138337. 128502.
-1  127507. 136838. 146501. 156222. 165241. 169456. 165498. 156323. 146552. 136828. 127419.
-2  124354. 132775. 141163. 148998. 154420. 156956. 154829. 149403. 141437. 132943. 124394.
-3  119611. 126957. 133311. 139449. 143417. 145167. 143844. 140064. 134449. 127245. 119736.
-4  113773. 120002. 125362. 129737. 132771. 134066. 133153. 130422. 126075. 120528. 114061.
-5  107239. 112297. 116697. 120172. 122609. 123597. 122968. 120894. 117497. 112984. 107720.
         -5      -4      -3      -2      -1       0       1       2       3       4       5
                                              g
```

Figure 10.3. Processing of binary images for computing the orientation (slope histogram) of the boundaries of grain profiles of the individual phases: (*a*) image of octagonally eroded pyroxene grains; (*b*) image of octagonally eroded scapolite grains; (*c*) image of all grains in the granulite after one octagonal erosion; (*d*) image of pyroxene grains after one square dilatation; (*e*) 1000 × 595-pixel image containing a 901 × 521-white pixel mask; (*f*) image of the intersection between the image in part *e* and the image in part *c*. This image has been shifted and intersected with the image in part *c* in order to obtain the geometrical covariance array in part *g* used for computing the slope histogram shown in Figure 10.4*e*.

shown in Figure 10.4A; that for octagonally eroded pyroxenes (PYOE1) is shown in Figure 10.4B. Both show two maxima that are stronger than the corresponding patterns for the scapolite grains shown in Figures 10.4C and 10.4D. The strongest orientation pattern was obtained for the octagonally eroded grains of the entire granulite (AXOE1 in Fig. 10.4E for one erosion and AXOE2 in Fig. 10.4F for two erosions). For comparison, the orientation pattern of the binary image of a circle is shown in Figure 10.4H after appropriate scaling. In this diagram the variation of the frequencies is restricted to a narrow interval, very small in comparison to the estimated perimeter of the circle itself. The rose diagram computations in RODIA are based on the estimation of the total particle perimeter from the measurement of the "intercepts" of the particles in all directions. The intercept is a measure of particle profile elongation in a particular direction, and it can be easily computed from the geometrical covariance values for small shifts (e.g., 4 to 10 pixels in length in the program). Except for Figure 10.4H, the rose

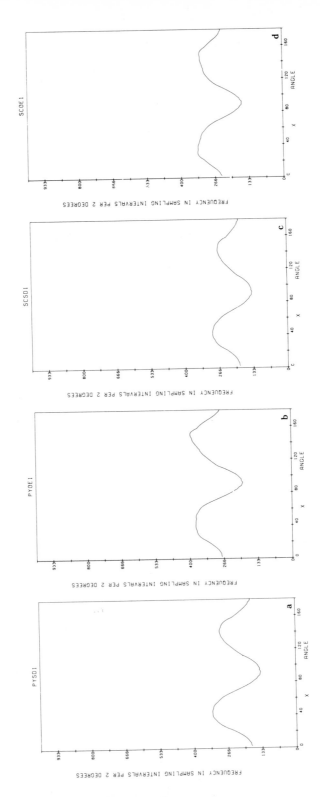

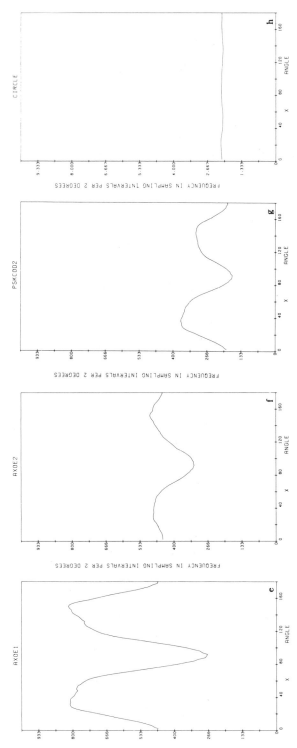

Figure 10.4. Slope histograms or rose diagrams computed for the pyroxene and scapolite grain-profile images. On vertical axes, frequencies 0 to 1000 are plotted versus angles 0° to 180° on horizontal axes: 0° is for the horizontal left-to-right direction, 90° is for the vertical top-to-bottom direction, and 180° degrees is for horizontal right-to-left direction. (a) Rose diagram for pyroxene after one square dilatation, (b) for pyroxene after one octagonal erosion; (c) for scapolite after one square dilatation, (d) for scapolite after one octagonal erosion; (e) for all crystal profiles after one octagonal erosion and (f) after two octagonal erosions; (g) for the pyroxene "skeletons" complemented after two octagonal dilatations; (h) for a circular-shaped profile (frequencies between 0 and 10).

143

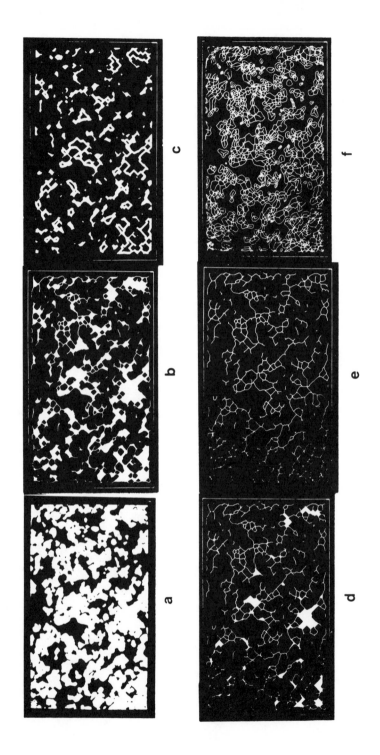

144

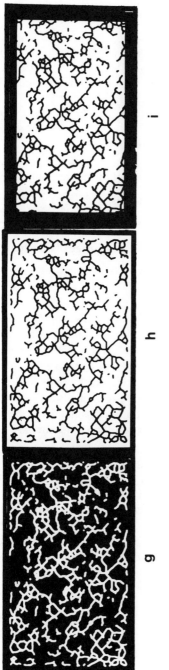

g   h   i

Figure 10.5. Experiments for computing the binary images of the "skeletons" of pyroxene profiles by line thinning, and for deriving the skeletons of orientation pattern: (*a*) binary image of the pyroxene profiles after one square closing transformation (a dilatation followed by an erosion); (*b*) the image in part *a* after 10 thinning iterations; (*c*) image of the pixels that changed value after the tenth thinning iteration, which produced the image in part *b*; (*d*) the image in part *a* after 20 thinning iterations; (*e*) the completely thinned "skeletonized" image in part *a* after 41 thinning iterations; (*f*) the union of the skeleton image in part *e* with the image of pyroxene boundaries; (*g*) the image of pyroxene skeletons after two octagonal dilatations; (*h*) the complement of the image of delatated skeletons in part *g*; (*i*) image of the intersection between the image in part *h* and the mask shown in Figure 10.3*e*, for computing the slope histogram shown in Figure 10.4*g*.

145

diagram for the circle, all diagrams in Figure 10.4 have been scaled similarly between the values of 0 and 1000. Within this range fall all the estimated perimeters in the various experiments.

The orientation diagram for the "skeletons" of pyroxene grain profiles shown in Figure 10.4g, which also shows two peaks of different magnitudes in the same directions, was obtained from images that were as shown in Figure 10.5. The procedure employed can be described in steps as follows:

1. square "closing" of the image of pyroxene grain profiles by a 3 × 3 black pixel structuring element;

2. expansion of the closed binary image;

3. line thinning to compute the "backbones" of the clusters of grains;

4. compression of the binary image of skeletons;

5. octagonal dilatation of the binary image of skeletons of pyroxene clusters;

6. computation of the complement of the dilated image of skeletons;

7. computation of the intersection between the complemented image of dilatated skeletons and a binary mask.

All the rose diagrams in Figure 10.4 reflect the anisotropy of the granulite texture as orientation of the grain profile boundaries. The pyroxene grains form chains of profiles in the 45° direction, where the contacts between adjacent pyroxene profiles are mostly perpendicular to that direction (i.e., in the 135° direction). It is very likely that the 45° direction corresponds to a lithological layering and to a faint gneissosity. Such anisotropy is also measured for the scapolite grains, but it is less visible. Although our eyes cannot detect the anisotropy for the entire granulite fabric, the overall pattern of grain-boundary anisotropy is the strongest.

A possible reason why the "Markovity" property in this fabric was observed by Whitten, Dacey, and Thompson (1975), and why various randomness tests gave negative results in the study by Kretz (1969), is the anisotropic pattern of the grain-profile boundaries. The possibility should be considered that such property of the grain-profile boundaries occurs in "ideal granites" (Vistelius and coworkers, 1966-1972; Vistelius and Harbaugh, 1980). This consideration would justify an analysis of thin sections of such ideal granites by the image-processing methods here exemplified.

COMPUTATION OF THE GEOMETRICAL COVARIANCE OF GRAIN PROFILES OF THE GRANULITIC FABRIC IN DIFFERENT DIRECTIONS

Is there in the images of the two major components of the granulite, pyroxene and scapolite profiles, a preferential distribution of the areas of the profiles in

146

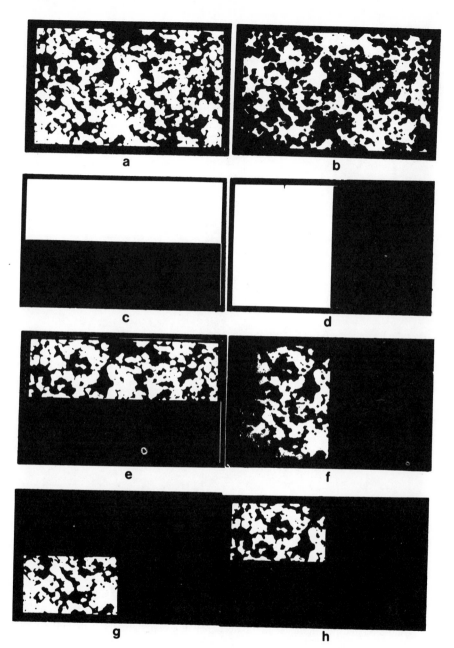

Figure 10.6. Preprocessing for computing the geometrical convariance of grain profiles in the granulite: (*a*) image of pyrixene profiles; (*b*) image of scapolite profiles; (*c*) upper half mask (1000 × 295 pixels); (*d*) left half mask (500 × 595 pixels); (*e*) intersection between the image of pyroxene profiles in part *a* and the image in part *c*; (*f*) intersection between the image in part *a* and the image in part *d*; (*g*) intersection between the image in part *a* and the non-overlapping part of the images in parts *c* and *d*; (*h*) intersection between the image in part *a* and the overlapping part of the images in parts *c* and *d*. Similar intersections were computed with the image of scapolite profiles in part *b*.

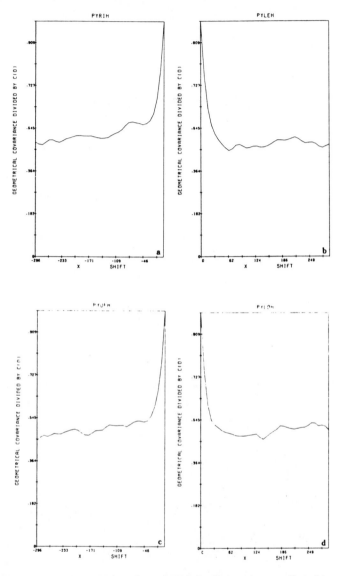

Figure 10.7. Linear geometrical covariances for the pyroxene profiles of the granulite computed in different directions: (*a*) horizontal direction, right to left; (*b*) horizontal direction, left to right; (*c*) vertical direction, top to bottom; (*d*) vertical direction, bottom to top; (*e*) 135° direction, bottom to top; (*f*) 45° direction, top to bottom.

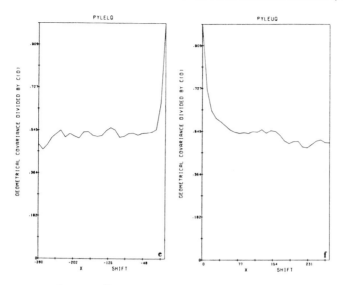

Figure 10.7. *(continued)*

the different directions? This question can be answered by computing the geometrical covariance for distances of pixels corresponding to several hundreds of individual shifts.

The preprocessing of the binary images of the two phases, shown in Figures 10.6A and 10.6B, before the computation of the covariance, is as follows. Masks for the upper (Fig. 10.6C) and lower half of the images and for the left (Fig. 10.6D) and right halves were computed, as well as for the upper left and lower left quarters. The intersections of the pyroxene image with some of these binary masks are shown in Figures 10.6E to 10.6H. Each masked image was intersected with the entire original image after being shifted in the directions in which it could travel without its edges reaching the edge of the latter. Linear geometrical autocovariances for 0°, 90° and 180° were obtained for both pyroxene and scapolite (see Figs. 10.7 and 10.8). For the pyroxene, the covariances for 45° and for 135° were computed, as shown in Figures 10.7E and 10.7F. Figures 10.7 and 10.8 are plots of the areas of overlap (intersection) relative to the initial area for shift equal to 0 in the two directions, which has the value of 1 and is plotted on the vertical axis. On the horizontal axis the shifts are plotted in pixels. It can be seen that the area decreases stabilize within shifts of 60 pixels. For this reason the two-dimensional geometrical covariance arrays shown in Figure 10.9 (see pages **152-153**) were obtained for unit shifts of five pixels between −60 and +60 pixels in all directions.

The values in Figure 10.9A for pyroxene, and in Figure 10.9B for scapolite, are the raw intersection pixel counts. The 801 × 381-black pixel mask used for the images to be translated is not shown here. Contours are added by hand for intervals corresponding to 70, 55, 52, and 40 percent of the area for shift (0,0).

149

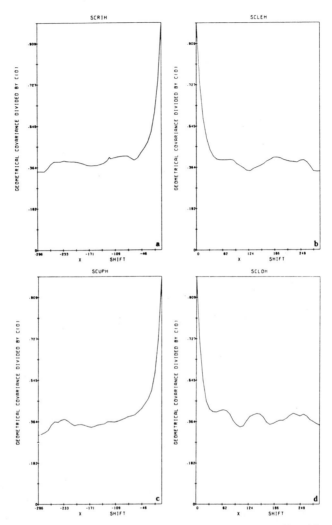

Figure 10.8. Linear geometrical covariances for the scapolite profiles of the granulite in different directions: *(a)* horizontal direction, right to left; *(b)* horizontal direction, left to right; *(c)* vertical direction, top to bottom; *(d)* vertical direction, bottom to top.

The contours show mildly anisotropic patterns, elongated in the 45° direction, which correspond to the direction of the gneissosity.

In Figure 10.9A, the anisotropy of the pyroxene profiles is indicated by the contours for all values; in Figure 10.9B, the anisotropy of the scapolite profiles is indicated by only the 40-percent contour.

A clearer, simplified expression of these anisotropies in the distribution of the areas of the grain profiles can be seen in Figures 10.9C and 10.9D. There the spacing between the covariance values are the same in the two directions.

The same contour lines of Figures 10.9A and 10.9B have been redrafted in Figures 10.9C and 10.9D for the covariance patterns of pyroxene and scapolite, respectively. Here we can observe not only the area distribution anisotropy but also the fact that the slopes of the geometrical covariance function is steeper for scapolite, which is a more disperse phase, than for pyroxene, a more clustered phase.

These results, therefore, support the suggestion of a high degree of homogeneity in grain-profile area distribution in the texture for both major components of the crystalline fabric. This would suggest that shape anisotropy is the strongest property of the fabric.

CONCLUDING REMARKS

The study of the granulite serves as an introduction to the types of problems that should be studied by image analysis and also to the kinds of methods that are applicable. Some of these apects have not yet been considered formally in petrology. The visual side of the present approach helps with method development and data analysis. The next chapter discusses a more formal geometrical statistical approach, which is part of stereology.

REFERENCES

Agterberg, F. P., 1979, Algorithm to Estimate the Frequency Values of Rose Disgrams for Boundaries of Map Features, *Comput. Geosci.* **5:**215-230.

Agterberg, F. P., and A. G. Fabbri, 1978, Statistical Treatment of Tectonic and Mineral Deposit Data, *Global Tectonics Metallog.* **1:**16-28.

Agterberg, F. P., C. F. Chung, S. R. Divi, K. E. Eade, and A. G. Fabbri, 1981, *Preliminary Geomathematical Analysis of Geological, Mineral Occurrence, Data, Southern District of Keewatin, Northwest Territories, Geological Survey of Canada,* Open File 718, 31p.

Kretz, R., 1969, On the Spatial Distribution of Crystals in Rocks, *Lithos* **2:**39-65.

Underwood, E. E., 1970, *Quantitative Stereology,* Addison-Wesley, Reading, Mass., 274p.

Vistelius, A. B., 1966, Genesis of the Mt. Belaya Granodiorite, Kamchatka (An Experiment in Stochastic Modeling), *Acad. Sci. (USSR) Proc.* **167:**48-50 (Akad. Nauk. (SSSR) Doklady **167:**1115-1118).

Vistelius, A. B., 1972, Ideal Granite and Its Properties: I. The Stochastic Model, *Internat. Assoc. Math. Geol. Jour.* **4:**89-102.

Vistelius, A. B., and Faas, A. A., 1971, The Probability Properties of Sequences of Grains of Quartz, Potassium Feldspar, and Plagioclase in Magmagranites, *Acad. Sci. (USSR) Proc.* **198:**170-172 (*Akad. Nauk. (SSSR) Doklady* **198:**925-928).

Vistelius, A. B., and J. W. Harbaugh, 1980, Granitic Rocks of the Yosemite Valley and an Ideal Granite Model, *J. Math. Geol.* **12:**1-24.

Vistelius, A. B., and Romanova, M., 1972, The Concept of Ideal Granites and Its Use in Petrographical Survey and Search Tasks (on Material from the Mesozoic Granitoids of Northeast Asia), in Vistelius, A. B., ed., *Ideal Granites - Issue* I (in Russian), Nauka, Press, Leningrad, Lab. Math. Geol., Acad. Sci., p. 4-47.

Whitten, E. H. T., M. F. Dacey, and K. Thompson, 1975, Markovian Grain Relationships of a Grenville Granulite, *Am. J. Sci.* **275:**1164-1182.

a

b

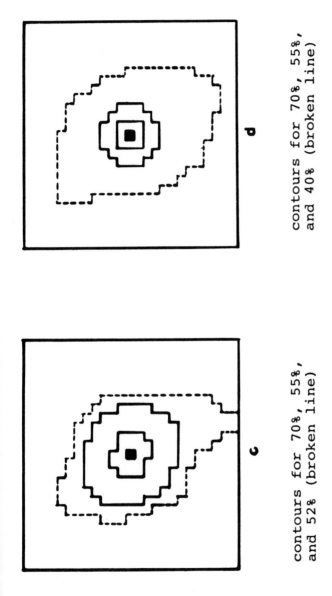

contours for 70%, 55%,
and 52% (broken line)

c

contours for 70%, 55%,
and 40% (broken line)

d

Figure 10.9. Two-dimensional geometrical covariances for pyroxene (*a*) and for scapolite (*b*) profiles of the granulite for 12 shifts of five pixels in all directions. Hand contours for 70, 55, and 52% (broken line) of the central value (no shift) were drawn in *a*; contours for 70, 55, and 40% (broken line) for the scapolite were drawn in *b*. The patterns of contours are distorted because the spacings of the values in the two directions are different. Simplified drawings of the contours for 70, 55, and 52% (broken line) of the geometrical covariance of pyroxene profiles are shown in *c* and of the contours for 70, 55, and 40% (broken line) for the scapolite profiles in *d*.

CHAPTER 11

Experiments on the Characterization of Metamorphic Textures from a Micrograph of an Amphibolite

This chapter describes some experiments on texture characterization from a thin section of a metamorphic rock and compares the results obtained by different interactive approaches. One approach uses a Quantimet 720 image analyzer, a specialized hardwired instrument working in real time but with limited processing capabilities. The other approach is slower computationally but has general programming capabilities. As described by Fabbri (1980), it uses a Modcomp II minicomputer (64K of 16 bit words of memory), a flying spot scanner, and image-processing software written in FORTRAN with only few machine-dependent routines.

The Quantimet 720 image analyzer is faster, but it has a limited computational power; it is ideal for routine work. On the Modcomp II general-purpose minicomputer it is possible to perform any type of computation, although certain tasks can require much computing time, which could be costly. The two approaches are compared in an attempt to determine which measures or transformations more efficiently describe the particular material analyzed. The experiments in this chapter were initially described by Fabbri and Masounave (1980).

This chapter has been adapted from A. G. Fabbri and J. Masounave, 1980, Experiments on the Characterization of Metamorphic Textures from a Microphotograph of an Amphibolite (a paper presented at 5th International Congress for Stereology, Salzburg, Austria, September 4-8, 1979), *J. Microscopy* **121**:111-117; *Mikroskopie* **37**:339-344.

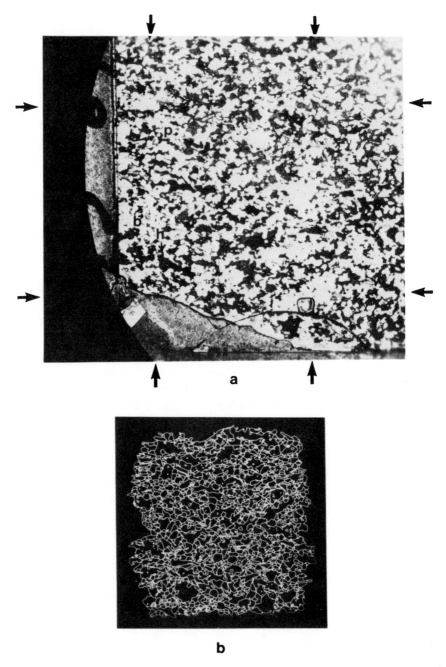

Figure 11.1. (*a*) Portion of the thin section of amphibolite that was studied. The study area is indicated by arrows; the trace of the foliation plane is horizontal. The letters h, b, and p indicate hornblende, biotite, and plagioclase crystals, respectively. (*b*) The boundary tracing of all grains in the study area.

THE MATERIAL ANALYZED

The material analyzed is a thin section of a Precambrian amphibolite from Sparrow Lake (Muskoka area, Ontario); see Figure 11.1A for a photograph of a portion of the thin section. The sample was collected in Precambrian terrains of the Grenville Province of the Canadian Shield. The orientation of the macroscopically visible foliation and lineation was recorded in field, and the thin section was cut perpendicularly to both the foliation plane and the lineation direction. The six different crystalline phases in the section are hornblende (50.19%), plagioclase (40.39%), biotite (5.48%), sphene (3.24%), apatite (0.68%), and zircon (0.02%) Frequently, the grains are not distinct, and uniformly colored; crystals of the same phase are in contact so that the boundaries between individual grains are not easily detected. For these reasons it is not possible to achieve satisfactory phase recognition and extraction with automatic scanning devices that use only optical techniques for phase identification.

PREPARATION OF IMAGE MATERIAL

An 8 × 8-mm portion of the thin section was projected onto tracing film, and the boundaries of over 2000 crystals, whose profiles were contained entirely within the square, were drafted at a magnification of 100×. The transparency (Fig. 11.1B) was photographed, using 35-mm film, for scanning by a flying spot scanner. A gray-level image was produced according to a hexagonal raster configuration of 900 × 1016 pixels. The image was thresholded to produce a binary image of the line drawing of crystal boundaries to be studied. After minor interactive editing in places of poor resolution, and hexagonal line thinning of the boundaries, the extraction of the different phases as separate binary images was obtained by component phase labeling as exemplified by Fabbri and Kasvand (1980). Some of the processing steps for this image are shown in Figure 11.2.

Two crystalline phases—biotite and hornblende—were selected for textural measurements. These phases display marked shape complexity and distinct orientation properties. Two black-and-white ink drawings of the phases, shown in Figures 11.3A and 11.3B, have been obtained from the original draft for the Quantimet 720 experiments. To allow for the separation of adjacent grains, care was taken that the white gaps between them were wide enough to be detectable on that instrument. Displayed in Figure 11.4 are portions of the thinned boundary image (A), the images of biotite (B) and hornblende (C), and the image of vertically eroded hornblende (D). The plots are pseudohexagonal; they were produced on a Versatec dot matrix printer from the Modcomp II computer system.

THEORETICAL BACKGROUND

The experiments have been restricted to three selected characteristics: mean traverse, orientation, and structure.

157

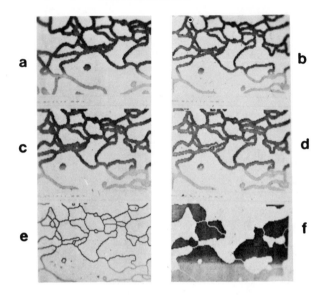

Figure 11.2. Digitization of boundary tracing and extraction of the binary image of amphiboles: *(a)* part of a binary boundary image obtained on the Modcomp II computer by thresholding an image after square raster scanning of the negative in Figure 11.1*b*, on a flying spot scanner; *(b)* thresholding after hexagonal scanning; *(c)* thresholding after hexagonal scanning at a gray-level range different from *b*; *(d)* edited cersión of the image in *c*; *(e)* line thinning of the image in *d*; *(f)* extraction of the binary image of hornblende crystal profiles from the image in *e*, after component labeling and interactive phase labeling.

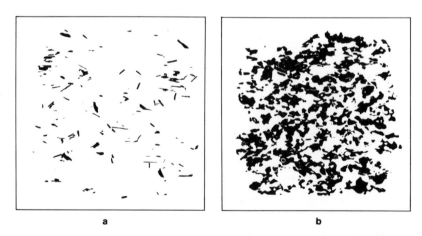

Figure 11.3. Ink drawings of biotite *(a)* and hornblende *(b)* crystals obtained from the boundary line tracing for the Quantimet 720 experiments. The foliation trace in these drawings is horizontal.

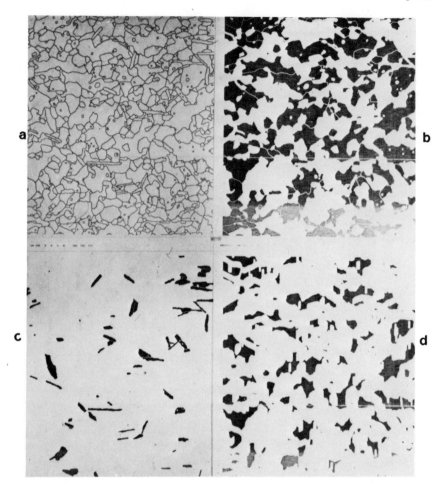

Figure 11.4. (*a*) Part of the hexagonally digitized binary image of boundaries of crystal profiles from the thin section of amphibolite: (*b*) binary image of hornblende grains, and (*c*) of biotite grains; (*d*) vertical linear downward erosion of the hornblende image by a structuring element 15 pixels long.

The usual way to measure the mean traverse of particles is based on the size distribution of their intercepts. The intercept function $I(l)$ is used, which is a function of the total number of chords detected after an erosion of length l. This function is used to compute the mean traverse "in number": all the traverses across the particles in a given direction have the same weight.

The mean traverse in number in a given direction can be defined as

$$m_1 = 1/I(0) \int_0^\infty l(\partial I/\partial l)dl \qquad (11.1)$$

159

where l is the length of the linear structuring element used for the erosion and $I(0)$ is the total number of intercepts with noneroded profiles. The subscript 1 indicates a one-dimensional measure.

The experimental problem with equation (11.1) comes from the instability of the $I(l)$ function on an image analyzer where the image is scanned continuously. The total number of intercepts, as well as the perimeter, is sensitive to the instrumental noise. It is possible partially to avoid this problem by using the $P(l)$ function, which also is the probability that a segment or a structuring element of length l is included in the grains. It can be shown (Matheron, 1967) that the $I(l)$ function is proportional to the first derivative of the $P(l)$ function. Thus equation (11.1) becomes

$$m_1 = -1/P'(0)\int_0^\infty l(\partial^2 P/\partial l^2)dl \qquad (11.2)$$

which gives, by integration by parts:

$$m_1 = -P'(0)/P(0) = \beta_1 \qquad (11.3)$$

where $P'(0)$ is the first derivative, for $l = 0$, of the $P(l)$ function. The $P(l)$ function, measured with the area function after an erosion, is less sensitive to the instrumental noise than the $I(l)$ function. We want to compare the results from equations (11.1) and (11.3) to decide which of the two is to be preferred.

EXPERIMENTAL RESULTS

The mean traverse m_1 is obtained by computing all the intercepts for the eroded particles for all possible successive erosions of length l in a given direction α. β_1 can be computed by producing a few successive erosions to obtain $P(1)$, $P(2)$, ... In practice, two or three erosions are sufficient.

The $P(l)$ function is measured by the Quantimet 720 with the area function after an erosion of length l; on the Modcomp II it is measured by a routine performing hexagonal erosions and dilatations of binary "compressed" images with structuring elements of any size and shape.

To take care of the frame bias during the erosion and the covariance measurements that will be described for the Quantimet and the Modcomp II computer, all the computations were obtained within a rectangular area smaller than and fully contained within the image analyzed, also after the transformations. Therefore, the $P(l)$ function measured is not sensitive to the frame bias: the reduction of the frame during the erosions or during the translations.

The relationship between the two measures in microns, m_1 and β_1, is shown in Figure 11.5A for hornblende and biotite for different angles of direction α, in degrees. The diagram suggests that β_1 is a more reliable measure of mean

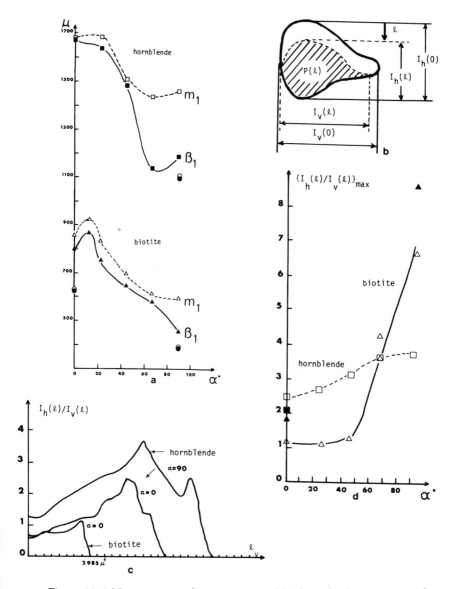

Figure 11.5. Mean traverse and orientation tests. (*a*) Relationship between m_1 and β_1, two measures of mean traverse in microns ($\simeq$) for biotite and hornblende profiles, and their variation with the angle of direction $\pm$. Open (m_1) and solid (β_1) circles indicate Modcomp II computer measurements; triangles and squares indicate Quantimet 720 measurements. (*b*) Definition of horizontal (h) and vertical (v) intercepts for a convex object at the origin ($I(0)$) and after a vertical erosion of length l ($I(l)$). $P(l)$ is the area of the eroded object. (*c*) Variations of the ratios of horizontal over vertical intercepts with successive linear erosions of length l, for different values of the angle of direction α. The data are from the Quantimet. (*d*) Variations of the maxima of the ratios of horizontal over vertical intercept for eroded biotite (triangles) and hornblende (squares) with the angle of direction α. Solid symbols are measurements from the Modcomp II computer; open symbols are from the Quantimet.

161

traverse on the Quantimet. Because of the absence of noise, on the Modcomp II both measures have the same values in different directions.

The Modcomp II measurements confirm that m_1 and $ß_1$ have values that are very close. This could not be done on the Quantimet. Clearly the mean traverse $ß_1$ is a more desirable measure in both instruments, where it is computed quickly with only three successive erosions. In the case of $ß_1$ both approaches are satisfactorily fast.

In Figure 11.5A we can observe an orientation effect: the mean traverse values vary with α. A phase can be considered oriented if it contains elongated particles and the orientation of each particle is appreciably related to that of its neighboring particles. For both biotite and hornblende, most grains are elongated in the horizontal, $\alpha = 0$, direction. Because the areal proportion of biotite (5.48%) is far less than that of hornblende (50.19%), from the diagram of Figure 11.5A we can say that biotite is more strongly oriented: the variation range of its mean traverse is proportionally greater.

More information can be obtained from measuring the ratios of horizontal to vertical intercepts as functions of a linear erosion by a segment l, in different directions. Figure 11.5B shows the vertical and horizontal intercepts at the origin (i.e., for $l = 0$), $I_v(0)$ and $I_h(0)$, and after a vertical linear erosion by l, $I_v(l)$ and $I_h(l)$. Figure 11.5C shows the ratio, computed on the Quantimet, of intercept functions versus the cumulative length of horizontal linear erosion. This ratio for different directions, in cases of convex grains with no holes (inclusions), describes an orientation effect in mixtures of large and small particles: large particles are more strongly oriented. The maxima of the curves, in Figure 11.5C, emphasizes maximum anisotropy of the particles that have their largest diameter in the direction of erosion. The variation of the maximum ratios for different directions is a good measure for characterizing the orientation pattern of biotite and hornblende, as shown in Figure 11.5D. The smooth pattern for hornblende is partly a result of the pronounced concavity of the crystals, the presence of holes, the elongated shape of these holes, and their positions within the host crystals. This can be seen in the erosion pattern of hornblende shown in Figure 11.4D.

The correspondence between Quantimet and Modcomp II measurements is good (see Fig. 11.5D). The performance of the Quantimet is, however, much more rapid (minutes versus hours) for sequences of tens of erosions in different directions, such as those shown in Figure 11.5C for the Quantimet only.

The structure effect occurs when the individual particles of a phase present some repetitive character in their distribution in some direction. This effect may depend on the orientation and is expressed in quantitative form by the covariance function, which represents the probability that both extremities of a linear segment oriented in a given direction are included in the grains belonging to the same phase. This function is measured on the Quantimet by a linear correlator module, which works in the horizontal direction only, and from left to right. Measurements in all directions can be made by rotating the image. On the Modcomp II computer this function is programmed in the two dimensions: two identical copies of a binary image are shifted along columns and rows and then compared (or correlated) after the translation. The result is a coefficient expressed in pixels of overlap, from which the geometric covariance can be easily computed.

The values of the geometrical covariance, expressed as proportions of pixels of overlap within the area analyzed, are plotted (vertical axis) against the shift lengths in microns (horizontal axis) in Figure 11.6. Individual subdivisions of shift length correspond to five pixels. For the two images analyzed, no strong effect of structure can be seen. The texture can be interpreted as the result of a homogenizing metamorphic process indicated by microfolding and shearing of foliation planes and recrystallization along the different planes. The metamorphic processes have obliterated any periodical pattern of grain distribution that may have existed before, within a distance of $9000\ \mu$.

Linear covariance measurements can be computed within one or two minutes for each direction on the Quantimet. On the Modcomp II computer they are at least ten times slower. One advantage of the latter approach is the possibility of computing the covariance in any direction without changing the setting of the image.

CONCLUDING REMARKS

Microscopic morphological features of amphibolites usually are not described quantitatively. The phases studied have mean traverses, orientation, and distribution patterns that can be characterized by the two approaches used in this chapter. Therefore, it will be possible to develop quantitative models for nucleation and metamorphic crystallization processes also based on the morphological and textural properties of rock fabrics. This approach has not yet been properly investigated.

The extraction of image information from the material analyzed, which was not attempted before, is cumbersome and technically complex: it is for this reason that little is known about the quantitative characterization of metamorphic textures from thin sections.

More work will be needed to speed up further the computing time required for the binary transformations programmed on the Modcomp II computer, especially for lengthy sequences of transformations that are produced in real time by the Quantimet 720 image analyzer. The image-processing approach is useful for testing measures that are affected by high noise on the Quantimet, and in particular for extending the processing to more generalized structuring elements and also beyond the analysis of binary patterns.

In the near future, hardwired image analyzers and medium- to large-size computers will merge or be interfaced, thus closing the gap between the two approaches. More modern versions of the image analyzers, such as the Quantimet 900, can store binary images in memory for further processing. The results of the simple experiments performed in this chapter suggest how software and hardware can be used in further developments of image-analyzing systems, and

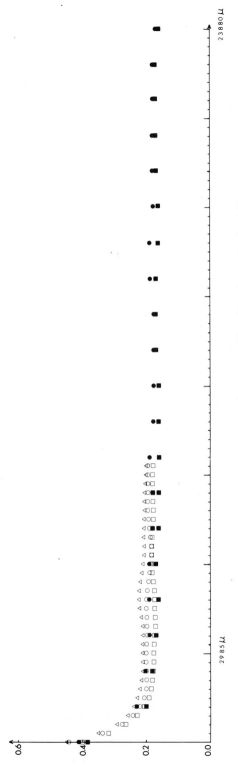

Figure 11.6. Covariogramme for the image of hornblende crystals for different angles of direction. Open symbols indicate Quantimet 720 measurements; solid symbols indicate Modcomp II measurements. Different symbol shapes indicate the different directions of measurement: circles, 0°; squares, 90°; triangles, 45°.

how new programs need to be written. It is hoped that a software-controlled image analyzer will be designed to be completely programmable and able to capture and store both binary and gray-level images for further processing.

Because of its working speed, an image analyzer is particularly well suited for routine work. However, it can compute a limited repertoire of transformations. The experiments described here are a first step in searching for efficient measurements or transformations by comparing two instruments. Nevertheless, a software-dependent system is not restricted in the choice of transformations so that new programs can continuously be designed for better characterizing image material.

REFERENCES

Fabbri, A. G., 1980, GIAPP: Geological Image Analysis Program Package for Estimating Geometrical Probabilities, *Comput. Geosci.* **6:**153-161.

Fabbri, A. G., and Kasvand, T., 1980, A Picture Processing Approach to Stereological Problems (a paper presented at 5th International Congress for Stereology, Salzburg, Austria, September 4-8, 1979), *Mikroskopie* **37:**431-436.

Fabbri, A. G., and J. Masounave, 1980, Experiments on the Characterization of Metamorphic Textures from a Microphotograph of an Amphibolite (a paper presented at 5th International Congress for Stereology, Salzburg, Austria, September 4-8, 1979) *J. Microscopy* **121:**111-117; *Mikroskopie* **37:**339-344.

Matheron, G., 1967, *Eléments pour une théorie des milieux poreux*, Masson et Cie., Paris, 166p.

CHAPTER 12

Petrology and Textures

REVIEW OF STUDIES OF ROCK TEXTURES FOR THE ANALYSIS OF MICROSCOPIC SECTIONS

During the past two decades several attempts have been made to explain textural patterns in igneous and metamorphic rocks in terms of free energy at grain boundaries. The following chapter is a review of contributions that are related to the study performed in this book (Chapters 10 and 11), in that they are concerned with the analysis of rocks in thin or polished sections. Most of the approaches in these contributions provide the background for more extensive applications, particularly if the assumption can be made of having at our disposal a digitized image of the section to be studied. Although this review is not complete, the topics reviewed are thought to be a sufficiently representative sample of geological studies.

In a study of grain contacts in granitic rocks, Rogers and Bogy (1958) assumed that if the distribution of different minerals is random and uniform, the percentage of contact area of a mineral that is in contact with any other mineral should be proportional to the modal percentage of the latter in the rock. In this situation it can be assumed that the former mineral did not affect the development of neighboring grains. They observed that potassium feldspar-potassium feldspar and plagioclase-plagioclase contacts were less frequent than expected in a random type of distribution, and concluded that the growth of one feldpar probably prevented the nucleation of a similar crystal. An alternative interpretation is that a growing potassium feldspar grain causes rapid nucleation of the other feldspar lattices in its neighborhood and simply incorporates these into its own structure rather than permit their formation as separate grains (Rogers, 1961). The information they used was in terms of what they called the "geometric means of the ratio: the percentage of contact length of a mineral A in contact with mineral B/modal percentage of mineral B." The modal data were obtained by point counting traverses and the number of times each possible contact was crossed. The value of this ratio for the potassium felspar-potassium feldspar contacts in the granites that they analyzed was 0.45.

Mahan and Rogers (1968) used the same method in a study of grain contacts in some high-grade metamorphic rocks. They computed a ratio of 0.83 for potassium feldspar-potassium feldspar contacts. The relatively high ratio in metamorphic rocks was interpreted as indicative of crystallization of the potassium feldspar in a comparatively solid medium in which a growing grain could probably have less effect on its environment and on nucleation in its neighborhood than it would have in a melt.

Whitfield, Rogers, and McEwen (1959) established definite relationships among textural properties and modal compositions of some granitic rocks. De Vore (1959) proposed a model in which relative grain sizes, grain distribution, extent of various grain associations, and nature of grain contacts are interpreted as an expression of a minimum interfacial free energy of the mineral assemblage. It is a nucleation-crystal growth model with a significant influence of the interfacial free energy on the crystallization of grain assemblages. According to the model there seems to be a definite and valid free-energy relationship that, if permitted to operate, could determine grain distribution and place an upper limit on the modal composition of the phases in the system. Thus these macroscopic features of grain assemblages could be effectively treated in terms of thermodynamic equilibria.

Kretz (1966a) successfully applied a quantitative approach to the study of rock textures, and derived equations for the rate of nucleation from information on crystal size distributions. The study yielded information concerning the growth processes of the minerals and the processes of migration of their component elements through the rocks in which they occur. In another contribution published during the same year (Kretz, 1966b), he developed a line of investigation based on the interfacial angles at the contact between several grains and shapes of individual grains. He succeeded in demonstrating solid-state growth in metamorphic rocks describing that triple points in metamorphic rocks were in positions for theoretical minimum surface energy. Kretz concluded that several aspects of the shape of mineral grains in metamorphic rocks can be attributed to a local reduction or minimization of interfacial free energy.

Hobbs (1966) observed that the microstructure of some Australian tectonites that he analyzed by axial distribution analysis (AVA) was consistent with adjustment of grain boundaries under the influence of interfacial tension. The rules governing the relationships between grains have been carefully developed by metallurgists; they fall into two related categories: (*a*) topological rules that govern the ways in which the parts of an aggregate must be geometrically related to one another, and (*b*) rules involving the configuration that grains in an aggregate must adopt to be in equilibrium under the influence of the interfacial tension of grains. In a single-phase aggregate, the free energy associated with grain boundaries must tend to a minimum for equilibrium, so grain shapes and boundary relationships must alter to make this possible. In polyphase aggregates, the same condition holds, but adjustment may take place to produce different phases with a lower free energy.

Vance and Gilreath (1967) used the same ratio used by Rogers and Bogy (1958) in a study of phenocryst distribution patterns in some porphyritic igneous rocks. The contact relations between the phenocryst minerals studied reveal a moderate to strong general tendency for preferential synneusis (clustering dendency) of crystals of like minerals, which is balanced by an antipathetic relation between unlike minerals in the same rocks. Specific minerals seem to exhibit consistent patterns.

Kretz (1969) examined in detail the distribution of crystal and crystal boundaries in a thin section of pyroxene-scapolite-sphene granulite. The rock was homogeneous, and the crystals of pyroxene, scapolite, and phene seemed to be randomly distributed. He carried out several different statistical tests for homogeneity and randomness in the spatial distribution of the crystals. Different types of measurements were performed and different data were collected and analyzed from a single thin section. In particular the results of the application of one method, termed "line-transect method," also used by Vistelius (1966), led to the conclusion that the nucleation site of any crystal in the rock was independent of any other crystal in the neighborhood.

Similar statistical tests applied by Flinn (1969) on different rocks show that grains in gneisses, instead of being distributed at random, are arranged so that grains of the same phase tend not to occur in contact with each other. This arrangement was interpreted to arise from grain-boundary migration leading to the insertion of grains of one phase between pairs of like grains of other phases. The process was thought to be due to the fact that the interfacial energies of contact between like phases were greater than those between unlike phases. According to Flinn, unlike contacts in metamorphic rocks are statistically favored over like contacts at a high level of significance.

Erlich et al. (1972) observed that surface free energy, which is a function of surface area and grain neighborhoods, may be a dominant factor in petrogenesis. Textural variables such as grain shape and surface area of phase contacts per unit volume can provide information about reaction pathways and kinetics. In a study of the response of textural variables to metamorphic grade, they determined a progressive increase in the proportion of unlike grain-to-grain contact with increasing grade. This was interpreted as decreasing the surface free energy. Similarly an increase in average size of grains, by reducing surface area, should result in a decrease of surface energy. Erlich et al. concluded that a rock of a given composition affected by increasing metamorphism can conserve or decrease its surface free energy only by textural readjustment over a limited range. When this range is exceeded, chemical readjustment is probably triggered, producing new stable assemblages. It then seems that surface free energy may play a fundamental role in the development of these stable assemblages. Both shape and surface area of plagioclase grains seem more effective in defining the gradient than are standard compositional variables.

According to Byerly and Vogel (1973), who studied grain boundary processes and development of metamorphic plagioclase, the only explanations for Flinn's (1969) observations are the increased effect of impurities on lowering the surface energy for unlike boundaries and the higher activation energy for unlike boundaries due to differences in crystal structures and chemical composition, whereas like boundaries have low activation energies with the source material for diffusion being available in the immediate vicinity of the boundary. In other words, the low-energy like boundaries are more mobile than the high-energy unlike boundaries due to the differences in activation energies. Surface energy is a major factor in controlling the processes involved and is itself a function of the geometrical properties, the distribution of phases, and the distribution of impurities within the rock aggregate. According to Byerly and Vogel, grain-boundary migration and impurity segregation are related processes that occur with annealing of any crystalline aggregate; they can explain a major part of the variation of the plagioclases that occurs in metamorphic rocks.

Jen (1975) considered the spatial distribution of crystals in charnockitic granulites as a function of interfacial free energy. The spatial distribution of crystals may give information on nucleation sites and on the role of interfacial free energy. He determined that all three major fundamental types of spatial distributions may occur in a rock: clustered, regular, and random (also intermediate types such as antiregular and anticlustered). Because this occurs frequently, although in most of the rocks there are two instead of all three types, such a mixed mode of spatial distribution would be expected to be the rule rather than the exception. This suggests that an ideal situation of either random, regular, or clustered distribution of crystals in natural rocks is rare, and that total equilibrium is rarely achieved. Jen used point-sample and line-transect methods (Kretz, 1969) in his study, and measured an average of 769 transitions per section. The analyses of the transitions were made by computer programming and were based on chi-square tests.

Since the early 1960s grain boundaries have constantly been studied. Chadwick and Smith (1976) recently compiled papers describing the state of the art in this field. There the grain boundaries are described and attempts are made to explain their properties. According to one of the definitions in the book, a grain boundary is not simply a collection of dangling lattice sites, but rather a defect structure that may have some degree of regularity. Atomic arrangements at grain boundaries are investigated, as are the chemical properties. Much work is being done on computer simulation of grain boundaries and on the methods of characterizing grain boundaries in terms of structuring elements.

Whitten, Dacey, and Thompson (1975) and Whitten and Dacey (1975) studied sequences of mineral grain transitions along linear traverses across mutually perpendicular sets of serial sections of the same Grenville calc-silicate granulite studied by Kretz (1969) and also discussed in Chapter 10 of this book. Their rather exhaustive tests proved that the granulite possesses the Markov property (Vistelius, 1966, 1972; Vistelius and Faas, 1971; Vistelius and Romanova, 1972). This property they describe as one in which observed mineral grains are controlled by the composition of adjacent grains in a rock. Such property seems to reflect petrogenetically important but unidentified factors. These authors concluded that in the granulite under study, major constituents observed in traverses normal to the weak mineralogical layering show a distinct nonrandom distribution, which implies that contacts between grains of the same major mineral are always less abundant than would be expected if the grains were randomly distributed. The Markovian properties of the calc-silicate granulite, which is a metasedimentary rock, must result from a mineralogy produced by high-grade neocrystallization in the solid state. This conclusion is in contrast with a model used by Vistelius and coworkers (1966-1972) which implies that the same Markovian properties should be typical of "ideal" granites, representing primary crystallization produced by a melt.

Wadsworth (1975) used grain-transition probabilities among mineral phases to analyze variation in grain sequence among 60 samples distributed throughout the zoned Cornelia pluton in southwestern Arizona. He concluded that variations in the patterns of grain sequences that he recognized among the units of the pluton justified a completely new petrologic understanding of the Cornelia stock. In his study, Wadsworth structured grain-transition data according to the model of "embedded" Markov chains, which avoids the necessity of recognizing grain contacts among like species.

Vernon (1976) emphasized the importance of studying grain boundaries in both monophase and polyphase crystalline aggregates. Their characteristics can be related to the structural deformation and recrystallization that usually overlap in metamorphic processes.

In addition, mention should be made of few particular applications to a very practical problem, that of ore dressing, and that of digital and optical analysis of rock textures. Petruk (1976) applied quantitative mineralogical analysis of ores to ore dressing. He used size distributions of minerals of economic value to predict the optimum grind for liberating the minerals and the degree of liberation that would be obtained at this grind. He observed that measurements of the properties of free and locked mineral grains in mill products show a good correlation between predicted and actual liberation, therefore demonstrating the advantage of using an image analyzer (a Quantimet 720) in mineral beneficiation research.

Serra (1966) and Agterberg (1967) were among the first to analyze thin sections of rocks by coding the occurrence of the different grain profiles by overlaying regular grids on microphotographs. Their statistical analysis of the coded arrays of data was to illustrate new methods of studying spatial relationships between minerals or mineral groups and possibly contribute to petrological interpretations. Serra analyzed a thin section of Lorraine oölitic iron ore, by comparing experimental variograms computed in different directions for three minerals. Agterberg worked instead on thin sections of a gabbro from the Muskox layered intrusion in the district of MacKenzie (N.W. Territory), Canada. He computed the two-dimensional covariance function and a two-dimensional power spectrum of the thin-section coded data.

To complete this review, the following work is also of interest. Jeulin (1981) observed that the physical and chemical behavior of manufactured multiphase products such as ceramics, refractories, concrete, and composite materials depend widely on the mutual locations of phases in the products, what he termed "multiphase structure." He defined several morphological parameters and functions for the description of such structures. In particular, he proposed to perform phases vicinity measurements by determining the distribution of contacts of one phase with the other constituents of the material. A similar model was the ground for an algorithm proposed by Fabbri and Kasvand (1981) for a semiautomatic approach to the measurement of the "degree of memory" retained by the grain profiles, of a digitized crystalline fabric, along outward two-dimensional paths from any set of grains. By processing the digitized image, they computed distance-related functions for one-step, two-step,..., n-step adjacency. The procedure constructs tables of adjacency relationships, which are useful for testing spatial models of crystallization. The approach was extended further by Fabbri, Kasvand, and Masounave (1982) to the extraction of particular two-dimensional sequences of grains in a granulitic rock. Results of this work brought out hidden distributions and agglomeration structures of the grains in which reformation and crystallization history may not have obliterated an original structure precedent to the last crystallization events. The possibility is suggested that a so-called Markovian property observed by Whitten, Dacey, and Thompson (1975) in a sample of the same rock is the result of the geometry and shape of such clusters of different selected phases.

Miller, Reid, and Zuiderwyk (1982) developed the QEM*SEM image analyzer, an instrument based on a scanning electron microscope. The analyzer uses the signals obtained from an energy-dispersive x-ray detector and from a back-scattered electron detector to map the particular minerals present at each series of points covering the sample. Volume fractions and intercept length distributions for each mineral are obtained, and the contact surface area of each mineral with each of the others. From the latter, normalized association probabilities can be derived, which define randomness or specific mineral associations. According to the authors, recovery of minerals from ores depends on the distribution and association of the various minerals in the ore.

172

OTHER STUDIES RELATED TO POROSITY IN SEDIMENTARY ROCKS

Both the theory and the applications of mathematical morphology stem from the problem of characterizing porous media particularly in the field of oil reservoir rocks. A fundamental theoretical approach to the study of porous media is due to Matheron (1967), who analyzed the empirical notions of granulometries and permeability. In particular, it is the granulometry of the pores that allows one to account for the arrangement of the grains and the permeability of the media. The concept of erosion and dilatation, and of opening and closing of sets by structuring elements, should be used to describe in probabilistic terms the morphology of pores and grains in thin sections.

A description of porous media in reservoir rocks was made by Delfiner (1971, 1972) and by Delfiner, Etienne, and Fonk (1972). He applied Matheron's approach to a morphological study for the automatic measurements of thin sections of sandstones. The effect of injecting mercury in porous cavities can be compared to the effect of opening the image of the pores by circular structuring elements. According to Delfiner, "provided there are no (or few) bottlenecks to prevent communication between pores, the saturated volume coincides with the opening of the pores with respect to a sphere of a radius equal to the radius of curvature of the interface at that pressure" (1972, 212). For example, with increasing pressure, the mercury progressively touches the surface of the total pore space without being constrained by the narrow necks. Both the pores and the grains in thin sections of sandstones were studied in terms of size distribution with respect to opening transformations.

The distinction was made between the size distribution in number (of particles or grains) and the one weighted in measure (volume, area, or weight). The latter distribution is the steadier characteristic of a population of objects. In particular, Delfiner used the concept of the "star" of the grains (and of the pores) described as the mean area seen directly from the point of the grains (or of the pores). This star represents the mean surface of the grains if they are convex, and can be computed by linear structuring elements (linear granulometry). The star can be a very reliable and sensitive measure for estimating the average volume of the pore and the pore surface.

Serra (1982) has extended the stereological meaning of the star of the pores to the case in which particles or pores are identified and individually weighted (for example, by their area). Imagining the pores as a transparent medium and the grains as an opaque medium, a candle moved throughout the pore space will direct its light over an average area that defines the star of the pores. Serra considers the situation of a two-dimensional hexagonal raster, for which he provides an example of three structuring elements involved in the measure-weighted average of the area of the pores.

Preston and Davis (1976) proposed a method using Fourier optics and optical

173

data processing for the analysis of porous material. They studied photomicrographs of sandstones of units forming petroleum reservoirs. Experiments were performed using a simple optical system in which beam of coherent, collimated, monochromatic light created a Fourier transform of the image of a thin section when passing through it. In the photomicrographs the sand grains appeared black and the pores white. The grain-pore boundaries caused diffraction of the light. The angle of diffraction is a function of spatial dimensions of grains and pores in the image. Apparently, the interpretation of the diffraction patterns may be a far from trivial task.

CONCLUDING REMARKS

This review of geological studies on rock textures is here made because it represents a supporting background for the applications of image analysis described in Chapters 10 and 11, and also because it suggests that other applications can be considered which go beyond the imagination of a single specialist.

A systematic study of rock texture in thin section or polished section has not yet started, although it is certainly overdue. Much of the physical behavior of crystalline material depends on textural properties, that is, the morphology of the grains and their distribution and adjacency relationships.

REFERENCES

Agterberg, F. P., 1967, Computer Techniques in Geology, *Earth Sci. Rev.* **3**:47-77.

Byerly, G. R., and T. A. Vogel, 1973, Grain Boundary Processes and Development of Metamorphic Plagioclase, *Lithos* **6**:183-202.

Chadwick, A., and D. A. Smith, eds., 1976, *Grain Boundary Structure and Properties*, Academic Press, New York, 388p.

Delfiner, P., 1971, Etude morphologique des milieux poreux et automatisation des mesures en plaques minces, Ph.D. thesis, Unversity of Nancy.

Delfiner, P., 1972, A Generalization of the Concept of Size, *J. Microsc.* **95**:203-216.

Delfiner, P., J. Etienne, and J. M. Fonk, 1972, Application de l'analiseur de textures à l'étude morphologique de reseaux poreux en lame mince, *Rev. Inst. Français Pet.* **27**:535-558.

De Vore, G. W., 1959, Role of Minimum Interfacial Free Energy in Determining the Macroscopic Features of Mineral Assemblages, I. The Model, *J. Geol.* **67**:211-227.

Erlich, R., T. A. Vogel, B. Weinberg, D. C. Kamilli, G. Byerly, and H. Richter, 1972, Textural Variations in Petrogenetic Analyses, *Geol. Soc. Am. Bull.* **83**:665-676.

Fabbri, A. G., and T. Kasvand, 1981, Image Processing for the Detection of Two-Dimensional Markovian Properties as Functions of Distances from Crystal Profiles, in *Proc. 3rd European Symposium for Stereology, Ljubljana, Yugoslavia, June 22-26, 1981*, Stereologia Iugoslavica, v. 3, (suppl. 1), pp. 153-163.

Fabbri, A. G., T. Kasvand, and J. Masounave, 1982, Adjacency Relationships in Aggregates of Crystal Profiles, in *Proc. 6th Int. Conf. on Pattern Recognition*, Oct. 19-22, 1982, Munich, Germany, p. 1207; also in R. M. Haralick, ed., 1983, *Pictorial Data Analysis*, Proceedings of 1982 NATO Advanced Study Institute, Bonas, France, Aug. 1-12, 1982, Springer-Verlag, New York, v. F4, pp. 449-468.

Flinn, D., 1969, Grain Contacts in Crystalline Rocks, *Lithos* **3**:361-370.

Hobbs, B. E., 1966, Microfabric of Tectonites from the Wyangala Dam Area, New South Wales, *Geol. Soc. Am. Bull.* **77**:685-706.

Jen, L. S., 1975, Spacial Distribution of Crystals and Phase Equilibria in Charnockitic Granulites from the Adirondack Mountains, New York, Ph.D. thesis, University of Ottawa.

Jeulin, D., 1981, Mathematical Morphology and Multiphase Materials, in *Proceedings of the 3rd European Symposium for Stereology,* Ljubljana, Yugoslavia, June 22-26, 1981, Stereologia Iugoslavica, v. 3, (suppl. 1), pp. 265-286.

Kretz, R., 1966a, Grain Size Distribution for Certain Metamorphic Minerals in Relation to Nucleation and Growth, *J. Geol.* **74:**147-173.

Kretz, R., 1966b, Interpretation of the Shape of Mineral Grains in Metamorphic Rocks, *J. Petrol.* **7:**68-94.

Kretz, R., 1969, On the Spatial Distribution of Crystals in Rocks, *Lithos* **2:**39-65.

Mahan, S. M., and J. J. W. Rogers, 1968, A Study of Grain Contacts in Some High Grade Metamorphic Rocks, *Am. Mineral.* **53:**323-327.

Matheron, G., 1967, *Eléments pour une théorie des milieux poreux,* Masson et Cie., Paris, 166p.

Miller, P. R., A. F. Reid, and M. A. Zuiderwyk, 1982, QEM*SEM Image Analysis in the Determination of Modal Assays, Mineral Associations and Mineral Liberation, in *Canadian Institute of Mining and Metallurgy XIV Int. Mineral Processing Congress,* Toronto, October 17-23, 1982, Session VIII, Mineralogy Applied to Ore Dressing, pp. 3.1-3.20.

Preston, F. W., and J. C. Davis, 1976, Sedimentary Porous Materials as a Realization of Stochastic Processes, in *Random Processes in Geology,* D. R. Merriam, ed., Springer-Verlag, New York, pp. 63-86.

Rogers, J. J. W., 1961, Origin of Albite in Granitic Rocks, *Am. J. Sci.* **259:**186-193.

Rogers, J. J. W., and D. B. Bogy, 1958, A Study of Grain Contacts in Granitic Rocks, *Science* **127:**470-471.

Serra, J., 1966, Remarques sur une lame mince de minerai lorrain, *Bur. Recherces Géol. Miniéres Bull.* **6:**1-36.

Serra, J., 1982, *Image Analysis and Mathematical Morphology,* Academic Press, New York, 610p.

Vance, J. A., and J. P. Gilreath, 1967, The Effect of Synneusis on Phenocryst Distribution Patterns in Some Porphyritic Igneous Rocks, *Am. Mineral.* **52:**529-536.

Vernon, R. H., 1976, *Metamorphic Processes: Reactions and Microstructure Development,* Wiley, New York, 247p.

Vistelius, A. B., 1966, Genesis of the Mt. Belaya Granodiorite, Kamchatka (an Experiment in Stochastic Modeling), *Acad. Sci. (USSR) Proc.* **167:**48-50; *Akad. Nauk. SSSR Doklady* **167:**1115-1118.

Vistelius, A. B., and Faas, A., 1971, The Probability Properties of Sequences of Grains of Quartz, Potassium Feldsjar, and Plagioclase in Magmagranites, *Acad. Sci. (USSR) Proc.* **198:**170-172 (*Akad. Nauk. (SSSR) Doklady* **198:**925-928).

Vistelius, A. B., 1972, Ideal Granite and Its Properties: I. The Stochastic Model, *J. Math. Geol.* **4:**89-102.

Vistelius, A. B., and M. Romanova, 1972, The Concept of Ideal Granites and Its Use in Petrographical Survey and Search Tasks (on Material from the Mesozoic Granitoids of Northeast Asia), in *Ideal Granites—Issue I* (in Russian), A. B. Vistelius, ed., Nauka Press, Leningrad, pp. 46-47.

Wadsworth, W. B., 1975, Petrogenetic Significance of Grain-Transition Probabilities, Cornelia Pluton, Ajo, Arizona, *Geol. Soc. Am. Mem.* **142:**257-282.

Whitfield, J. M., J. J. W. Rogers, and M. C. McEwen, 1959, Relationships among Textural Properties and Modal Compositions of Some Granulitic Rocks, *Geochim. Cosmochim. Acta* **17:**272-285.

Whitten, E. H. T., and M. F. Dacey, 1975, On the Significance of Certain Markovian Features of Granite Textures, *J. Petrol.* **16:**429-453.

Whitten, E. H. T., M. F. Dacey, and K. Thompson, 1975, Markovian Grain Relationships of a Grenville Granulite, *Am. J. Sci.* **275:**1164-1182.

175

CHAPTER **13**

Toward Pattern
Recognition

We have been considering applications to geological or geophysical maps and to microscopic images of thin sections of rocks. We can classify those applications in terms of their different degree of uncertainties about what to measure from the material to be analyzed. A greater uncertainty is attached to geological map boundaries and to geophysical map contours than to grain-profile boundaries as seen under the microscope.

In the application that follows, we have a lesser degree of uncertainty about what to measure from a negative film of alpha-particle tracks, in order to count the tracks. To the human visual system the problem seems simple, and such trivial pattern-recognition tasks are taken for granted. For a computer, this task may require considerable computation, as we shall see, since every little detail requires particular procedures and computer programs.

AUTOMATIC COUNTING OF ALPHA-PARTICLE TRACKS FROM AUTORADIOGRAPHS OF RADIOACTIVE MINERALS

Image processing can be used to automatically separate and count randomly distributed segments from binary images of autoradiographs of the alpha-particle

This chapter has been adapted and expanded from T. Kasvand and A. G. Fabbri, 1978, Automatic Counting of Alpha-Particle Tracks from Autoradiographs of Radioactive Minerals, Proceedings of the Eighth Annual Automatic Imagery Pattern Recognition Symposium, R. A. Kirsch and R. N. Nagel, eds., April 3-4, 1978, National Bureau of Standards, Gaithersburg, Maryland, pp.11-26.

tracks emitted by radioactive minerals. Earlier versions of the approach have been described by Kasvand and Fabbri (1978) and by Fabbri and Kasvand (1981).

Uranium- or thorium-bearing minerals emit alpha particles, beta particles, and gammarays at rates proportional to the amount of U^{238}, U^{235}, and Th^{232} that they contain. A method for estimating the amount of radioactivity of those minerals consists of counting the number of alpha particles emitted by crystals that are located at the surface of polished sections or in thin sections. This is done by placing a film coated with alpha-particle–sensitive emulsion (5 to 200 μ thick) in contact with the surface of the section for a given length of time. The appropriate time must be empirically determined in each case to produce enough tracks for the estimation but not so many as to hinder visual separation and counting. For the present application a polished section with crystals of torbernite, a hydrated phosphate of copper and hexavalent uranium, has been used. Exposure times range between one and three hours. For a given exposure time, the ratio of the total number of alpha tracks to the total exposed crystal surface area is related to the surface alpha activity of the crystal, which in turn is a quantitative measure of uranium content. Quantitative aspects of alpha-particle patterns have been treated in the past by Yagoda (1949) and more recently by Rogers (1973).

The image analyzed here is a Kodak Contrast Process Ortho 4154 negative of a 130X enlargement of a portion of the cloud of tracks emitted by one crystal of torbernite in three hours. A positive print of the image is shown in Figure 13.1. Most tracks appear as straight, disjoint segments of different lengths and orientations. The darkness of the tracks varies irregularly along their length. A few tracks intersect and partly overlap in areas of higher density. Some tracks appear as darker, rounded blobs with diameters up to twice the width of the linear tracks. (The smaller and lighter dots are emulsion grains of the film; larger blobs are grains of dust.) This image represents a general case, encountered during routine visual counting; the density of tracks in the image is considered satisfactory for both visual and statistical estimates (S. Kaiman, pers. comm., 1978). In the framed area of Figure 13.1, 102 acceptable tracks and 3 unclassifiable blobs can be distinguished.

The task of counting the number of objects in a field, which seems relatively simple for the human eye, is in reality complex for a machine if a correct count is desired for an infinite variety of shapes, orientations, and overlaps.

In several applications, the overlapping individuals are removed from the field of analysis to eliminate the bias due to crossings (Cruttwell, 1974). Alternatively, manual interactions with image analyzers are used, such as light pen probes for isolating or eliminating specific particles from the screens of TV monitors (Jones, 1975).

The problem of separating and counting partly overlapping individuals from an image, for which the human eye and brain are so well suited, has not yet seen much progress. Recent tests made for counting asbestos fibers in air samples (Pavlidis and Steiglitz, 1976) and pulp fibers for paper manufacturing (Graminski and Kirsch, 1977) clearly exemplify both the difficulties of and the need for this

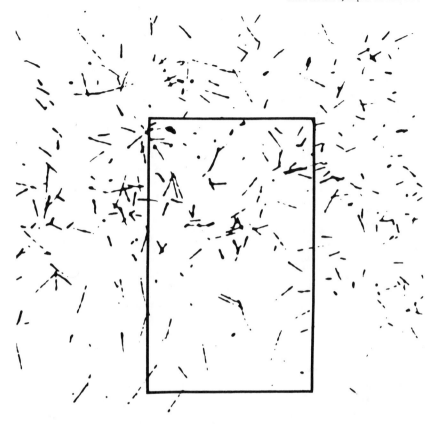

Figure 13.1. An enlarged positive print of a portion of alpha-particle tracks emitted by one crystal of torbernite. The framed area has been scanned and processed digitally.

task. The experiments described in this chapter attempt to achieve complete automation.

ALTERNATE APPROACHES TO THE PROBLEM OF ALPHA-PARTICLE TRACKS

The experiments can be carried out on a relatively small general-purpose computer, Modcomp II, with 64K of memory, 16 bit words, two 1-million word disks and two 800-bit-per-inch tapes. The input/output devices consist of a card reader, a dot matrix printer, a teletype, an online randomly addressable flying spot scanner, and a memory-type CRT with a special operator console. Such equipment is intended for general research on image processing rather than specifically for the alpha-particle track problem.

179

The equipment allows several possible approaches:

1. Because the online flying spot scanner is randomly addressable, the film transparency of the tracks could be used directly as a nearly homogeneous [to within the resolution of the digital-to analog (DA) converters of 13 bits] read-only memory to the computer. This approach has been used successfully on other problems (Kasvand et al., 1975; Kasvand, 1972). However, a randomly addressable flying spot scanner is not a generally available computer accessory.

2. There is a certain resemblance between the alpha-particle tracks and bubble-chamber photographs; specifically, both contain streaks (McIlwain, 1976). Consequently, some of the methods used in the analysis of bubble-chamber photographs could be used here, even if the specialized hardware is not available. From results obtained so far, it seems that similar line-enhancement algorithms may be required.

3. The least specialized and reasonably inexpensive picture-scanning device is a TV camera interfaced to some mass-storage device (tape, disk), usually via the computer. Thus a raster-scanned line-by-line image is captured for further computer processing. This is the approach that was selected. The film transparency was scanned line by line and the resultant gray levels stored on tape as a logical record per scan line. This produces an image that is far too large for the memory of a minicomputer.

The conflicting requirements of a large image size and small computer memory put certain constraints on the program structure. The use of "windows," or small areas of image data in memory, was considered difficult because a conglomeration of criss-crossing tracks could cover an area of arbitrary size. Thus there is no guarantee that the image of even one set of criss-crossing tracks could be contained wholly in the memory. Instead, the image is processed line by line, so that only one or a few lines are in the memory at a time. Thus a track is practically never "wholly visible" to the computer.

THE PROCESSING SEQUENCE

The fundamental choice to be made is whether to direct the programming effort specifically toward the alpha-particle track problem or to use a more generalized approach, in which the tracks are only a special case. The generalized approach, of course, is far more attractive from the research point of view, even though it results in rather extensive computations and is slower than an efficient specialized alternative. The danger with the specialized approach is that the opportunity is avoided to evaluate all the difficulties that might be encountered. Such difficulties, if they cannot be resolved, will be blamed then on poor picture quality.

In a research environment devoted to image processing, a fair set of applicable programs is already available, usually accompanied by an image-processing philosophy.

Consequently, the path of least resistance is to use whatever programs and methodologies are applicable, and to add new ones as required until a processing sequence has been established for pilot runs. Based on the experience gained and results obtained, the practical realization of a production setup can now be evaluated.

Opinions differ as to the actual algorithms to be used (Pavlidis and Steiglitz, 1976; McIlwain, 1976). It is unlikely that one unique best method exists for solving a reasonably complicated image-processing problem. The merit of a processing method is in its results. In the present instance, the following sequence of processing steps has been tried, and their successes and failures illustrated.

1. An enlarged film transparency (35 mm) is scanned (square raster) in a region where a characteristic set of tracks can be seen. An example is shown in Figure 13.1. The scanned pixels are written as one logical record per scan line: 161 records of 256 pixels.

2. The scanned gray-level image is high pass filtered to remove some scanner shading (Fig. 13.2) and the result thresholded to obtain a black-and-white or binary image (Fig. 13.3).

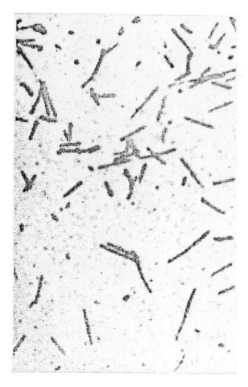

Figure 13.2. The scanned and high-pass-filtered image of the framed area in Figure 13.1. The image contains 161 rows of 256 pixels.

3. For every pixel on a track, the distance to its closest contour is computed to estimate the average thickness of each track.

4. The binary image is thinned to lines that are one element thick (Fig. 13.4) by using an eight-neighbor rule. The thinning simplifies slope computations but is mainly done to facilitate the location of junctions between crossing line elements and the listing of their connectivity. The black-and-white conversion and subsequent thinning result in some loss of information. It is believed, though, that this loss is outweighed by the resultant simplification of the problem. Specialized track sharpening at the gray-level stage will be required because some of the tracks tend to break.

5. All places on the lines where major changes in direction occur ("kinks") are located, and the black pixel at the kink is "removed" to break the lines (Fabbri and Kasvand, 1981a). A few examples of kink detection are illustrated in Figure 13.5.

6. The junction points between crossing tracks are detected and numbered sequentially. This cuts all crossing lines at the junctions. The resultant line

Figure 13.3. Binary representation of the scanned image in Figure 13.2. This black-and-white representation has been obtained by simple thresholding.

Figure 13.4. The thinned binary image obtained by thinning Figure 13.3 using an eight-neighbor rule. Computer analysis of the alpha-particle tracks is based on this image.

182

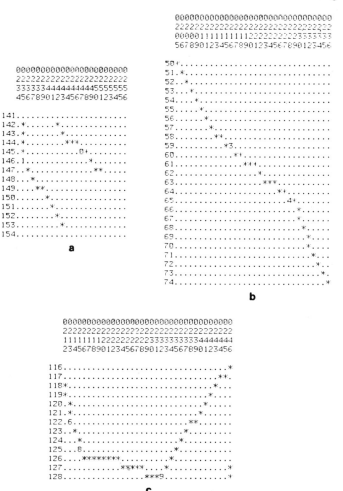

Figure 13.5. Detection of corner points or "kinks". (a) two segments separated into four tracks; (b) one segment separated into three tracks; (c) one complex segment separated into four tracks. Asterisks indicate black pixels, dots indicate white pixels, and numbers identify the detected kinks. In this illustration, as in the following ones, image x-coordinates (top rows) and y-coordinates (leftmost column) locate the pixels. Due to the unequal character and line spacing on a printer, all these images are stretched vertically.

elements will be termed *segments*. A segment may thus have: (a) no junction, in which case it is a free-standing line consisting of one or more alpha-particle tracks (if exactly parallel and end-to-end); (b) A single junction point; or (c) two junction points. A segment in instance (b) may consist of a single track connected at one end of a partial track. In instance (c), a segment may consist of a single track connected at ends with two tracks, or part of a track. Basically all combinations are possible.

7. Because the machine "sees" only one or a few scan lines at once — that is, only a few scan lines are in computer memory — each segment is given a unique label (serial number). At this point, the binary image contains serially numbered junctions and serially numbered segments, illustrated in Figure 13.6. The segments can be tabulated individually and their connectivity at a junction can be determined uniquely.

8. The local slopes of the line segments are computed, and these values are smoothed once or twice.

9. The individual segments meeting at every junction are checked to determine whether they can be connected. If a reasonably straight composite segment can be formed of two segments meeting at a junction, this connection is made, and the segment and junction tables are updated. This process is iterated until no further segment joining is possible.

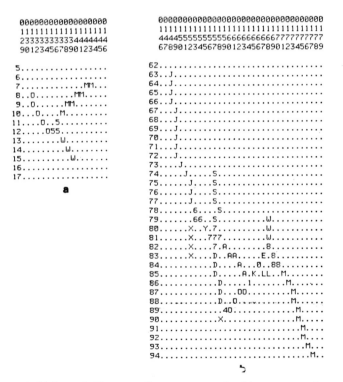

Figure 13.6. (*a*) A simple T-junction between two tracks; (*b*) a more complex track cluster. Letters symbolize segments, numerals indicate junction numbers, and dots represent the background.

184

10. As usual, a whole series of unexpected problems arise. The complicated junctions frequently consist of several neighboring junctions separated by segments that are only one or few pixels long. A typical case is shown in Figure 13.6B. The solution adopted at the moment is to amalgamate doubly connected short segments (i.e., segments connected at each end) and junctions. This makes the neighboring junctions merge, resulting in larger junction regions, but allows the connectivity algorithms, in step 9, to proceed past such imperfect junctions. The pixels forming the junction are added to the composite segments to obtain a reasonably accurate measure of track length.

11. Processing steps 9 and 10 are repeated until no further segment connectivity takes place. In most of the cases the resultant connectivity is comparable to human performance, given the somewhat degraded thinned binary image as a starting point. Several separated track clusters are shown in Figures 13.7A and 13.7B, and in Figures 13.8A and 13.8B. Of course, somewhat comical results may also occur (Figure 13.8C) even though the count may be correct.

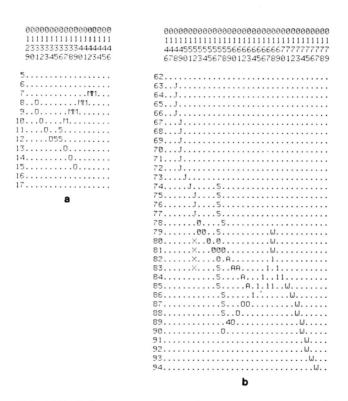

Figure 13.7. (*a*) Track cluster in Figure 13.6*a* after processing, giving two identified tracks labeled O and M; (*b*) track cluster in Figure 13.6*b* after processing.

185

```
00000000000000000000000000000000000000000
1111111111111111111111112222222222222
788888888889999999999990000000000111
901234567890123456789012345 6789012

11.......................Y............
12.......................Y............
13........XX...........................
14.........XX.........................
15...........X........................
16V..........XX........................
17.............XX......................
18.............XXX.....................
19..............XXX.......F...
20....G................9.....F....
21....G................9.99.FF.....
22....G...............999....9.......
23...9............FFF9...9.99........
24..99......FFFF....9..99..........
25..G..F.FFFF.......T.99...........
26..G...F...........T................
27..G.............TTT.................
28..G.........TTT....................
29..G.......TT......................
30..G...TTT........................
31..................................
                  a
```

```
0000000000000000000
1111111111111111111
4555555555556666666
901234567890123456

93..................
94..................
95......E...........
96A.....E...........
97......E...........
98......E....I......
99......E...I.......
100.......E.I.......
101........8.........
102.......8.........
103.......8.........
104......8..........
105.....I.E.........
106....I..E.........
107...I....E........
108...I.....E.......
109..I......E.......
110.........E.......
111.........E.......
112.........E.......
113..........E......
114..........E......
115..........E......
116...........E.....
117..................
118............,.....
         b
```

```
0000000000000000000000000000000000000000000
00000000000000000000000000000000000000000000
55566666666667777777777 8888888888999999
7890123456789012345678901234567890123456789012345

64..................................................
65..................................................
66..............................\..L.....
67.............................L.....
68............................L.....
69...........................4.....
70..........................P.L....
71.........................P...L...
72.........................P...L...
73.........................P...L...
74.........................P...L...
75........................P...L....
76........................5...L....
77.......................55.L......
78.......................L..5......
79......................L...P......
80......................L..P......
81......................L..P......
82......................L..P......
83.......................9.P........
84......................99..........
85.....................L..I........
86.....................L...I.......
87.....................L...I.......
88.....................L..I........
89U....................3.I.........
90U...................I.I.........
91....................I...........
92.................IIII.............
93.................III..............
94................II................
95...........III..................
96.........I......................
97........II......................
98......II........................
99....II..........................
100..II...........................
101.I.......M......................
                c
```

Figure 13.8. (a) Another resolved track cluster; (b) two tracks crossing at about 60°; (c) example of two close parallel tracks intersecting a third track at a high angle.

12. Except for certain problem areas to be discussed separately, the computational aspects of this problem are now no longer in the image-processing realm. A table of results is available that contains (a) the track number, that is, the final serial number assigned to the joined segments; (b) junction number(s) to which the track may be connected but further amalgamation cannot take

place; (c) the average slope of the track; (d) the length of the track, that is, its pixel count; (e) the average track thickness; (f) the average gray level along the thinned track (if wanted); (g) the x-y coordinates of the track center.

Additional information on track linearity, its "stringiness" (length/width ratio), average gray level, and so on, is or can be made available to form a decision space for recognizing tracks from other artifacts. Furthermore, the basis of an interactive display and verification stage now exists. A summary of the processing for automatic counting of alpha-particle tracks is shown in Figure 13.9.

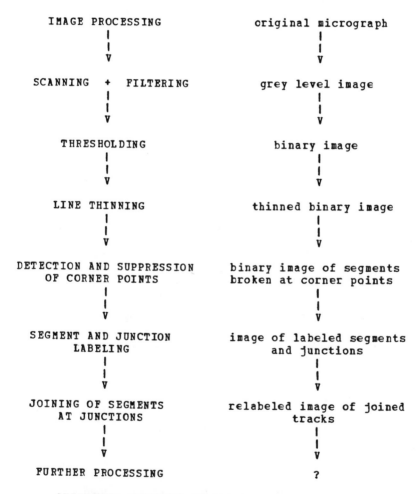

```
    IMAGE PROCESSING              original micrograph
           |                              |
           |                              |
           V                              V

  SCANNING  +  FILTERING            grey level image
           |                              |
           |                              |
           V                              V

        THRESHOLDING                  binary image
           |                              |
           |                              |
           V                              V

      LINE THINNING             thinned binary image
           |                              |
           |                              |
           V                              V

DETECTION AND SUPPRESSION      binary image of segments
    OF CORNER POINTS            broken at corner points
           |                              |
           |                              |
           V                              V

 SEGMENT AND JUNCTION          image of labeled segments
       LABELING                     and junctions
           |                              |
           |                              |
           V                              V

  JOINING OF SEGMENTS          relabeled image of joined
     AT JUNCTIONS                       tracks
           |                              |
           |                              |
           V                              V

   FURTHER PROCESSING                      ?
```

AUTOMATIC COUNTING OF ALPHA-PARTICLE TRACKS

Figure 13.9. Summary of processing steps for automatic counting of alpha-particle tracks. The left portion shows the flow of computational steps; the right lists the images generated in their processing sequence.

187

SOME REMAINING PROBLEM AREAS

As already mentioned, the gray-level image needs further filtering to prevent fragmentation of weak tracks in the binary image. At the same time, however, close parallel tracks or tracks crossing at a shallow angle should be converted to a proper image to allow the thinning algorithm to work properly. Figure 13.8C illustrates one such case. Two parallel and touching tracks may be distinguishable only by their nearly double widths.

Another problem is a track in the image plane crossed by another track nearly vertical to the image plane. In this situation the track contains a dot. Given proper film, the gray-level density at the crossing point can be higher than along the rest of the track. However, the gray-level density alone cannot be used because similar density values occur at two crossing tracks.

Two collinear tracks joined at their ends will form a track with higher density at the junction point. In appearance the resultant track would be similar to a long track crossed at nearly a right angle to the image plane by another track. Thus the track count is likely to be correct, but the interpretation will be incorrect. The large blobs of black and the small points are relatively easy to distinguish based on the distance to the closest contour and density measures.

CONCLUDING REMARKS

A rather lengthy sequence of processing algorithms is described and illustrated to solve the problem of counting alpha-particle tracks. From the practical point of view, the present solution may be too research-oriented and lengthy to be of direct practical use. Considerable simplification of programs is possible, however, and many processing steps can be collected into a single program.

The programs allow overlapping curved tracks to be studied, connected, measured, and counted. The constraint of straightness posed by the alpha-particle tracks is used only in two places, namely, at the segment-joining stage where additional decision parameters related to curvature have been zeroed, and at the segment linearity-checking stage.

Satisfactory overall counting correspondence between human and machine have been achieved: the problem of track clusters, which poses one of the major difficulties for automatic counting, is basically solved. The complexity of the required mechanized solution illustrates the variety of procedures available when solving similar problems. To "understand" a picture, we ourselves apply a variety of methods to the problem, both with ease and without necessarily even being aware of all the complications. Obviously, if a machine is to rival human performance, each and every processing step has to be described logically.

The approach presented here can be applied to other types of images in which straight or curved line patterns occur, as was done by Kasvand (1978, 1979) for paper pulp fibers and for Chinese ideograms. One of its advantages is that the

188

output of these algorithms provides tables of labeled components, which can be used for statistical analysis of individual segments. The method represents an alternative to optical processing for quantification and selective extraction of linear features from maps or photographs in fracture trace analysis (Norman, Price, and Peters, 1977; Harnett and Barnett, 1977).

REFERENCES

Cruttwell, I. A., 1974, Pattern Recognition by Automatic Image Analysis, *Microscope* **22**:27-37.

Fabbri, A. G., and T. Kasvand, 1981, Applications at the Interface between Pattern Recognition and Geology, in *Sciences de la Terre, Serie Informatique Geologique*, no. 15, pp. 87-111.

Graminski, E. L., and R. A. Kirsch, 1977, Image Analysis in Paper Manufacturing, in *Proceedings of the IEEE Computer Society Conference on Pattern Recognition and Image Processing*, June 6-8, 1977, Troy, New York, pp. 137-143.

Harnett, P. R., and M. E. Barnett, 1977, Optical Rose Diagrams for Lineament Analysis, *Trans. Inst. Mining Metall.* **86**:B102-B106.

Jones, E. J., 1975, Practical Aspects of Counting Asbestos on the Millipore MC, *Microscope* **23**:93-101.

Kasvand, T., 1972, Experiments with an Online Picture Language, in *Frontiers of Pattern Recognition*, S. Watanabe, ed., Academic Press, New York, pp. 223-264.

Kasvand, T., 1978, Experiments on Automatic Extraction of Paper Pulp Fibers, in *Proceedings of the 4th International Joint Conference on Pattern Recognition*, Kyoto, Japan, November 7-10, 1978, pp. 958-960.

Kasvand, T., 1979, Ideogram Segmentation and Recognition, in *Proceedings of the International Conference on Cybernetics and Society*, October 1979, IEEE, pp. 674-678.

Kasvand, T., and A. G. Fabbri, 1978, Automatic Counting of Alpha-Particle Tracks from Autoradiographs of Radioactive Minerals, *Proceedings of the Eighth Annual Automatic Imagery Pattern Recognition Symposium*, R. A. Kirsch and R. N. Nagel, eds., April 3-4, 1978, National Bureau of Standards, Gaithersburg, Maryland, pp. 11-26.

Kasvand, T., P. Hamill, K. C. Bora, and G. Douglas, 1975, Experimental Online Karyotyping at the National Research Council of Canada, in *Automation of Cytogenetics*, M. L. Mendelson, ed., Asilomar Workshop, Pacific Grove, Cal., Nov. 30-Dec. 2, 1975, pp. 96-109.

McIlwain, R. L., Jr., 1976, Image Processing in High Energy Physics, in *Topics in Applied Physics*, vol. 11, A. Rosefeld, ed., Springer-Verlag, New York, 1976, pp. 151-207.

Norman, J. W., N. J. Price, and E. R. Peters, 1977, Photogeological Fracture Trace Study of Controls of Kimberlite Intrusion in Lesotho Basalts, *Trans. Inst. Mining Metall.* **86**:B78-B90.

Pavlidis, T., and K. Steiglitz, 1976, The Automatic Counting of Asbestos Fibers in Air Samples (a paper presented at The Third Joint Conference on Pattern Recognition, Nov. 8-11, 1976, Coronado, Cal.), *IEEE Comput. Soc. Proc.* **76CH1140-3C**:789-792.

Rogers, A. W., 1973, *Techniques of Autoradiography*, Elsevier, Amsterdam, 372p.

Yagoda, H., 1949, *Radioactive Measurements with Nuclear Emulsions*, Wiley, New York, 356p.

CHAPTER 14

Epilogue

This work spans three different aspects of geological image processing: the philosophical approach of image processing in geology, the programming of the image-processing software, and several applications to geological problems, including one pattern-recognition experiment.

What are the consequences of this study? Is image processing really usable in geology? Should we expect more in the future, or should we consider this experiment as an isolated, peculiar venture, destined to remain in the records of methods searching for an application?

Of course, a complete answer to such questions cannot be provided by this work alone. Nevertheless, the applications performed suggest that many geological problems can be formulated with models and can be presented in forms very similar to image-processing problems. A computer's capability to complement human vision in some quantitative tasks and to act as an interactive helper is being currently exploited by many geologists. New and powerful techniques for the definition and subsequent recognition of desirable "environments" in economic geology are needed because of the demands for resource estimation.

Just as an automobile represents an extension of our senses—we can actually feel a boulder under the wheel of a car almost as if we touched it—or as the microscope gives us the opportunity to penetrate the microcosmos, so computer processing can tell us something about the properties of crystalline fabrics. It can quantitatively characterize those properties in a systematic manner that is often entirely beyond our physical capabilities. A machine can detect a foliation that we cannot see although we suspect it exists. It can tell us how much more strongly one

191

set of crystal profiles is oriented than another set; it can also count objects, faithfully and persistently, relieving us of tedious and error-prone mechanical tasks. Of course, this machine does require our guidance for it to perform with any intelligence.

A geologist should not be afraid to use a novel technology, because as soon as the suspicion of the machine and its rules has been overcome, he has a faithful assistant for expanding his work in many conventional studies but also in some new systematic aspects of his research work. Although the applications described in this book have the main purpose of enriching the available software with algorithms that could not be imagined without practice, they also serve to familiarize the geologist with the image-processing approach by way of example. It is my hope that this commitment has contributed in communicating with the geological community and perhaps with other researchers in the geosciences.

A Computer
System Dedicated
To Interactive
Image Processing

In the development of the techniques described in this book, the electronic equipment used was a computer system intended for general research in image processing. The system forms part of the Computer Graphics Laboratory of the Electrical Engineering Division of the National Research Council of Canada in Ottawa. The electronic equipment in the laboratory can be classified broadly into the following five classes: (a) computer and peripherals, (b) interactive devices, (c) display devices, (d) digitization devices, and (e) recording devices. In this broad classification, however, we must keep in mind that the instruments grouped under one class may perform the roles of some other classes.

THE COMPUTER AND PERIPHERALS

A small general-purpose computer, Modcomp II, with a read/write memory of 64 K (thousand) 16 bit words, two 1 million-word disks and two 800-bpi (bits per inch) tape drives, was used for processing images. The peripheral input/output (I/O) devices consist of a card reader, a Versatec dot matrix printer, a teletype, an online random addressable flying spot scanner, a memory-type Tektronix 611 CRT (cathode ray tube) with a special operator console, and a Summagraphics graphic tablet digitizer. The computer and the peripherals are diagrammatically shown in Figure A.1. The interconnections of the devices in the laboratory are illustrated in Figure A.2.

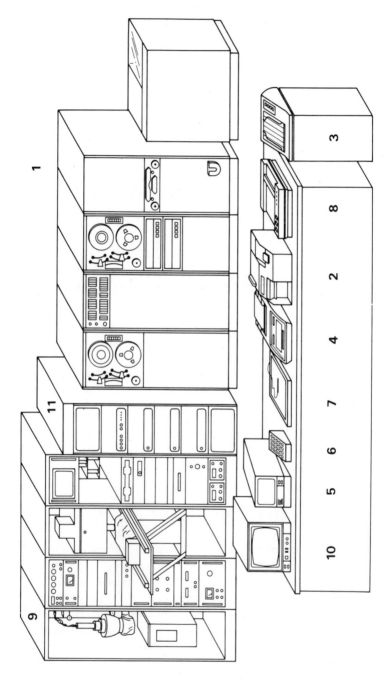

Figure A.1. The Modcomp II computer and the peripheral equipment dedicated to general-purpose image processing at the Computer Graphics Laboratory: (1) Modcomp II computer; (2) card reader; (3) Versatec dot matrix printer; (4) teletype; (5) memory-type Tektronix 611 CRT, with (6) special operator command console; (7) Summagraphics graphic tablet digitizer; (8) X-Y Hewlett Packard 7221A plotter; (9) online random addressable flying spot scanner; (10) Conrac color CRT, connected to (11) Norpak image buffer.

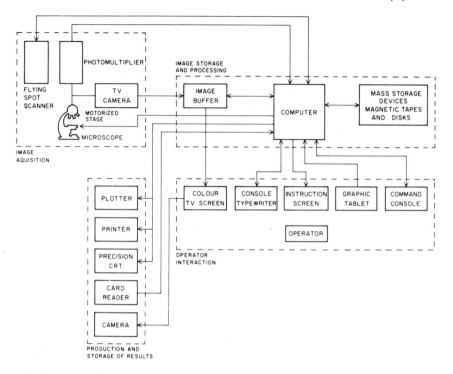

Figure A.2. The interconnections of the various devices in the Computer Graphics Laboratory.

INTERACTION DEVICES

The interaction equipment consists of a teletype and a special operator console connected to the Tektronix 611. The operator console is a push-button board consisting of 18 push-buttons with indicator lights, six potentiometers with indicator lights for analog signal input, and six additional indicator lights for general purposes. The illumination level of the console light is controlled via FORTRAN-callable routines.

DISPLAY DEVICES

The display equipment consists of the Tektronix 611 memory-type CRT, which allows the display of four gray levels for an array of 4096 × 4096 points, a Versatec dot matrix printer/plotter for black-and-white only, for an array of 1024 × 746 dots per 8½ × 11-inch page, and a Conrac color television that displays arrays of 640 × 512 points.

195

DIGITIZATION DEVICES

The digitization equipment consists of an online random addressable flying spot scanner, manufactured by Litton, for 35-mm transparencies, which has a resolution of 4096 × 4096 addressable positions over a 25 × 25-mm area; and of a Summagraphics graphic tablet with a resolution of 1500 × 1500 positions over a 15 × 15-inch area. FORTRAN-callable routines control both digitizers.

This particular flying spot scanner, which can be used for both input and output of pictorial data, is essentially a CRT that allows access to any point in a two-dimensional data plane. In its "hard-copy" mode, it can generate 64 levels of light intensity at each point and can handle a maximum matrix size of 4096 × 4096 points. Input comes from a transparent film, and output can be obtained using a standard Polaroid camera. Because the flying spot scanner operates at computer speed, it can be used as a random-access analog memory.

Lines can be traced manually on the graphic tablet with a specialized stylus. The tablet and the stylus system is connected to the computer, which receives and stores the x-y coordinates of the stylus while the operator traces the lines on the tablet.

RECORDING DEVICES

The recording equipment used for picture data consists of two magnetic tape drives (many pictures are stored serially and in bulk), two disk drives (scratch, used to store serially only a few pictures), and of the various display units. In the computer memory only part of one or more images can be stored for truly random access. Also a special memory was added to the system: called *image buffer* or *frame buffer*, it can store an entire color picture of 640 × 512 pixels, for random access. This buffer was built by Norpak Ltd. of Ottawa.

B

GIAPP:
Geological Image
Analysis
Program Package

An interactive system of FORTRAN programs, termed Geological Image Analysis Program Package (GIAPP), has been designed as part of multidisciplinary efforts in the fields of pattern recognition and mathematical geology (Fabbri, 1980, 1981). Several methods of image processing, mathematical morphology and stereology, have been brought together in programming the package. As a tool for the study of geological images, it is adaptable to any general-purpose minicomputer: it is written in FORTRAN with few machine-dependent routines. A version of the package that was adapted to a multiuser CYBER computer uses only a Tektronix graphic terminal for image display and interactive commands and messages. In essence, GIAPP offers the possibility to do image processing without having to use an image analyzer.

GIAPP was programmed to handle images of maps or micrographs as matrices of numbers for computation; to quantify all the information that they contain; to describe the morphology of each map unit or microscopic phase; to study the spatial relationships between map units; and to provide inputs to multivariate analysis automatically.

SUMMARY OF GIAPP'S CAPABILITIES

Table B.1 summarizes the capabilities of GIAPP. Some of the terminology in the table is that of Rosenfeld and Kak (1976), who provide a comprehensive overview of digital image processing. The programs have been developed first on a small

TABLE B.1
General Description of the Capabilities of GIAPP

General Features	Processing of Noncompressed Images	Processing of Binary Compressed Images
file searching, copying erasing, and reviewing	smoothing, filtering, thresholding, and graphical displays	logical operations on or between binary images
expansion and compression of binary data	component labeling of binary images	Minkowski operations by means of programmable structuring elements of any shape
handling of commentaries added to image data sets by each processing routine	line thinning	
	phase labeling	two-dimensional auto- and cross-correlations
processing of binary, labeled, and gray-level images for both square and hexagonal rasters	junction detection	point and vector displays of binary images
	extraction of binary data from labeled images	editing of binary images
output of image data on magnetic tape for data transfer	digital displays	creation of binary images from data on cards or computed from inter- active commands (masks, grids, tests)
	other special-purpose algorithms	
creation of binary compressed images of boundaries from a graphic tablet	false color displays	
scanning by means of a flying spot scanner		

computer dedicated to general image processing, as described in Appendix A. Later the package has been converted to a CDC CYBER 74 computer, which allows 70 octal K of memory of 60 bit words for interactive processing and the access to several disks. In this second version a Tektronix 4014/1 video graphic terminal and a hard-copy unit connected to it are the only communication and display devices available.

DATA MANAGEMENT

GIAPP was designed to make maximum use of a limited computer memory: a read/write memory of 64 decimal K words of 16 bits. On the Modcomp II computer, the operating system uses 12K of memory. The remaining 52K of memory are reserved for (a) the controlling part of the program (here called SUPERMASTER), (b) the integer and the real common areas, and (c) one group of programs conveniently grouped into an overlay (here called MASTER). This last part (c) of the memory is occupied in turns by each particular overlay, or MASTER, during an interactive session.

198

Initially the programs are stored on a magnetic tape and have to be transferred to a disk before each run. Upon starting, a system of overlays is entered, from which there is no exit until the processing work is completed or until it is interrupted manually by activating an external interrupt switch. Each overlay is accessed and exited interactively by teletype commands. Optionally the controlling commands can be issued by computer cards, which are read by the card reader.

An associated image file-management system for image processing (Fabbri, Kasvand, and Stray, 1978) provides information about image data, such as headings with names and dimensions, and comments on the history of processing of each image. By means of a simplified conversational language, logical operations between images or pixel neighborhood transformations can be computed by algorithms designed for arrays of up to 1024×1024 bits (1080×1080 on the CYBER computer). From an operational point of view, images are treated as data files. The programs allow selection of options and calls to various subprograms, which generate new images as output files. The latter can be used as input data for further processing. Plotting subprograms for graphic display on the video terminal can be called, as well as routines that modify the image data into input files for other display devices. On the CYBER version of GIAPP, hard copies of the text of the interactive conversation and of the binary images displayed can be obtained by a Tektronix 4631 hard-copy unit.

The processing of binary images in which each pixel corresponds to one bit of a computer word requires bit manipulation. All character and bit manipulation in the system is carried out using routines programmed in assembler language. Disk files can be accessed directly. Specialized file-handling routines were written to allow sequential file searching of the image data files on magnetic tapes.

For the system a data set is a structured file of image data. As exemplified in Figure B.1, a data set consists of (a) a header record of 100 computer words, describing the image-sequential address, type, and dimension; (b) a variable number of comment records of 100 computer words each, containing the image name and all names of the processing algorithms to which the image was subjected, including the commentary of other images with which it was combined; (c) the image data as an $(m) \times (n)$ array of integer numbers, in which m is the record length and n the number of records; and (d) an end-of-file mark.

Several images can be stored sequentially in any of the storage devices, tapes or disks, where each set of images is preceded by a beginning-of-data record. The last data set is followed by an end-of-data record. Both records are of 100 computer words each similar to the header record mentioned earlier. The package allows up to five independent I/O devices, which are accessed by entering the numbers between 100 and 599. Each physically independent unit can serve either as an input or an output storage device. A physical unit cannot, however, be used in more than one role at the same time. Because storage is sequential, only the last data set can be erased; a new data set can be added only next to the last data set.

The system attempts to be as useful as possible. At the beginning of a run in which one or more existing image files are available, a procedure ("COOL START")

199

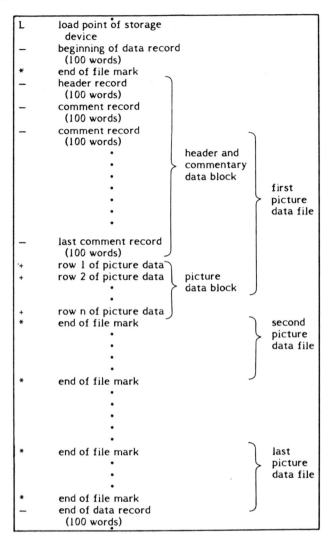

Figure B.1. Physical structure of sequentially organized image data in GIAPP.

reads the headers of all the existing files and creates a table in memory that contains all the information for file availability (input or output data sets) on all five storage devices. If during a processing step the user tries to read a nonexistent data set or tries to overwrite existing data, the system simply generates error messages and returns control to the user, without ruining the image data or aborting. As mentioned earlier, only the last image on a physical device can be erased.

THE CONVERSATIONAL STRUCTURE

GIAPP is a self-sufficient system that automatically handles storage and processing from several I/O devices (tapes or disks). This is possible via a simple conversational language, which is kept to a bare minimum by referring only to parameter addresses and to the corresponding values. The user is expected to specify three different classes of interaction: (a) to define I/O image numbers, (b) to define the running parameters for a given subprogram, and (c) to call by number the particular program. When the program completes execution, the control returns to the user. If an error is detected, an error message is generated and again the control returns to the user.

In the original Modcomp II version, a description of the programs and a list of their valid parameters and addresses and possible values must be used during the interaction. Some subprograms present various options to the user and require a response. These are explained in detail and in plain language by the programs during the interaction. The subprograms themselves are also called by numbers, which are explained in the description. In GIAPP it is the user's responsibility to know which label corresponds to which image and to draw a flowchart of the processing. That flowchart and a copy of the text of the interactive conversation will enable him to reconstruct the session at a later stage. On the subsequent CYBER version of GIAPP, a hierarchical system of menus and options has been inserted into the program code. The option to run the programs with the pre-existing minimal user interaction of the Modcomp II version is nevertheless retained because of its fast speed. An example of user interaction for a session of GIAPP on the CYBER computer is presented in Appendix D.

In general, GIAPP and its subprograms try to protect themselves against a careless user by checking as much input data as possible. The system is also "forgiving," however, since a detected error will only generate an error message (usually a program name followed by a number) and return control to the user, instead of returning control to the operating system of the computer. Thus, during a normal run the system does not allow an abort to occur.

GENERAL STRUCTURE OF THE PROCESSING ALGORITHMS

The optimization of processing speeds and the image data organization depend on the nature of the image-processing algorithms. Most algorithms, however, process images in a systematic manner, starting at the top left corner of an image and ending at the bottom right corner. Therefore, the images can be stored as large two-dimensional matrices in which each element and a certain number of its intermediate neighbors or the corresponding elements in other image matrices are used for computation. To minimize storage, only a strip across an image is stored in core. The width of the strip corresponds to the number of rows of input image data that are used by a given algorithm. The minimum strip consists of one row of pixels. Each row of image data represents a logical record. The image data are represented as matrices of integer or one-bit binary numbers of up to 1080×1080 elements. These are stored row by row in a sequential manner.

MACHINE-DEPENDENT ROUTINES

A few machine-dependent routines are used in GIAPP for bit manipulation. These routines are all FORTRAN-callable but are machine-specific. For this reason, they will have to be rewritten or substituted for the particular equipment on which GIAPP may be transferred and to which it may be converted.

The routines perform the following tasks for arrays of 64 computer words, that is, 1024 bits on the Modcomp II computer (18 words, or 1080 bits, on the CYBER computer):

1. Shift one bit to the left in the array, having a 0 binary bit loaded on the rightmost bit of the rightmost word.

2. Shift one bit to the right in the array, having a 0 binary bit loaded on the leftmost bit of the leftmost word.

3. Count the 1 binary bits in the array.

4. Set to 1 binary bit any given bit in the array.

5. Check the bit status (binary 0 or 1) of any given bit in the array.

ORGANIZATION OF THE PROCESSING ALGORITHMS OF GIAPP

The programs in GIAPP have been grouped into five classes:

I. Control of the flying spot scanner and processing of gray-level images;

II. digitization of line drawings by a graphic tablet, and editing of binary compressed images;

III. processing of binary images of boundaries to produce phase-labeled images and binary images of the different phases;

IV. processing and analysis of binary compressed images by means of structuring elements as defined in mathematical morphology; and

V. transfer of uncompressed image data outside the Modcomp II computer system.

Table B.2 summarizes the various tasks performed within the five classes of programs of GIAPP. Each class, termed a SUPERMASTER, consists of one or more overlays, termed MASTERS. The algorithms have been grouped, as shown in the table, to facilitate the interaction with processing steps, which are in logical and practical sequence.

TABLE B.2
Summary of the Functions Performed by the Groups of Programs in the Five Classes of SUPERMASTERS in GIAPP

Supermaster	Master	Functions
I	01	Scanning on a flying spot scanner, display of the results of the scans, creation of gray-level images, study and display of gray levels, thresholding of gray-level images, and creation of binary compressed images as labeled data sets for subsequent processing
	02	Selection of subpictures from large original images, point operations on pixels for gray-level images
II	03	Graphic-tablet programs for digitizing line drawings, creation of files of vectors from a graphic tablet, transformation of vectors into binary images, assemblage of subpictures (384 pixels × 384 pixels maximum) into large mosaics of binary compressed images of line drawings
	04	Assemblage of subpictures (512 pixels × 512 pixels maximum) into large mosaics of binary compressed images of line drawings
	05	Editing, via graphical displays, of binary compressed images
III	06	Line thinning of binary expanded images, component labeling of lines, junctions, or areas within contours
	07	Phase labeling via interactive displays, creation of phase-labeled images, extraction of binary compressed images of each phase
IV	08	Logical operations on or between binary compressed images, Minkowski operations or transformations by means of programmable structuring elements, displays of binary compressed images
	09	Auto- and cross-correlation of binary compressed images in two dimensions, creation of correlation images
V	10	Output of binary expanded images, labeled images, and gray-level images for transfer in different BCD formats

APPENDIX **C**

Parallel Processing Algorithms For Minkowski-Type Transformations Of Binary Images On A Minicomputer

A binary image has only two values of gray level: 0 or 1, which we can consider as white or black, respectively. It can be obtained directly by mechanical digitization or by thresholding a gray-level image at a given value: lower values become 0's, higher values become 1's, and vice versa.

Binary data can be compressed so that an ON-OFF bit status corresponds to each 0-1 pixel. This is done (a) to economize in memory storage, (b) to expedite I/O and computations, and (c) to utilize efficiently logical and bit shift operators, which exist on most computers.

As previously mentioned, in GIAPP all transformations are programmed in FORTRAN. Parallel processing, the contemporaneous transformation of the pixels of a binary image, is simulated for arrays of 1024 bits (1080 bits in the CYBER computer), corresponding to individual rows of image data. Logical operations between vectors of words (rows of image data), bit shift operations between vectors of 1024 bits (64 words), bit counting, bit status checking, and bit status-setting operations are combined into programs for the parallel processing of binary compressed images.

Four algorithms have been programmed in GIAPP for the following computations: (a) logical operations, (b) binary neighborhood transformations in which only matching or mismatching 1's are considered for both the square and the hexagonal images, (c) binary neighborhood transformations in which both the matching 0's and the matching 1's are considered, and (d) auto- and cross-correlation (geometrical covariance and cross-covariance).

A logical operation between two images is performed by comparing words in

205

both of them and generating, after each comparison, a word of output image. This is done sequentially, starting from the first word at the top left of the image and proceeding to the last word at the bottom right. These operations are extremely fast: they are programmed as assembler routines, which directly process strings of 64 words (1024 bits) in one single call from FORTRAN. As in any of the logical operations and transformations in GIAPP, the number of binary 1 bits is counted, by means of a fast bit-counter routine, as soon as a row of data is computed.

Autocorrelation (or geometric covariance) is performed by generating two copies of the same image (two different images are used for cross-covariance) and computing an .AND.'ing operation between two corresponding rows, after each shift (either to the right or to the left) for a given number of bits. Only the number of binary 1's in the computed row is counted and accumulated for each shift for each row. For horizontal shifts, bits are shifted; for vertical shifts, rows in one or the other of the two images are skipped before the process starts in synchrony. The output of the program is an array of counts of binary 1 coincident pixels for each shift position requested. The array can also be stored in the form of a gray-level image. Examples of correlation arrays can be seen in Figures 10.7 to 10.9.

Figure C.1 describes in detail the processing in "parallel," for computation (c) mentioned previously, of $(C=A \ominus B1)$ and $(D=A \oplus B1)$ of Figure 3.4. The structuring element is expanded into structuring-element windows to fit the image to be transformed. Both the image and the structuring-element windows are "padded" (i.e., filled with 16 bits of binary 0 values at both ends of each row). One row of padding is also added to the top and bottom of the image. Each row of structuring-element window, either shifted or unshifted, is compared with the corresponding image-window row containing as many rows as there are in the structuring-element window. The comparison with all rows of structuring-element windows for shifts 0, 1, and 2 are computed and one row of output image is generated, before shifting the image window down the image one row. The processing continues until the last row of output image is completed.

Both the eroded image and the dilatated image are displayed in the lower part of the illustration. The dimensions of the input and of the output images are 32 pixels × 12 pixels. Dots and 1's symbolize white and black pixels, respectively; this is so also for the structuring element and for the structuring element-windows, where the center pixel is underlined. Only 1's in the structuring element represented are used, in this case, for computing the two transformations. The dots represent the "don't care" positions in the neighborhoods. For detecting a match in a dilatation, a comparison — that is, the operation of .AND.'ing — is computed between two words at a time, each word consists of 16 pixels (in reality this is done for two sets of 64 words at a time): this represents a degree of "parallelism," a function of the word length (number of bits) of the computer. For an erosion, complementation, .NOT.'ing precedes .AND.'ing to detect mismatches. The processing can also be considered "pipelined," since the computations are structured row by row to optimize between processing time and available memory. Input/output access time is minimized because advantage is taken of the machine-dependent instruction set available for

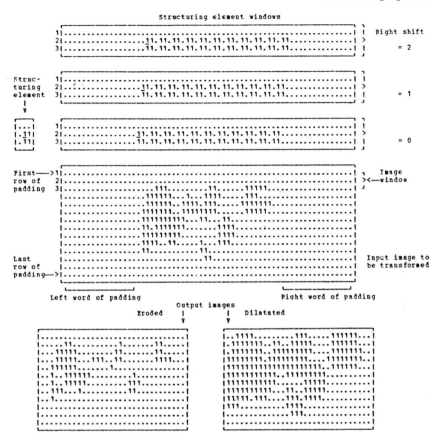

Figure C.1. Example of "parallel" processing of a binary compressed image in GIAPP.

logical operations and also for bit shift operations for arrays of 64 words (i. e., 1024 bits). The image and the transforms in Figure C.1 correspond to those illustrated on the left of the illustration in Figure 3.4.

Figure C.2 describes the processing, for computation (c) mentioned above, of $(F = A \ominus B3)$ of Figure 3.4. Most of what what was said for Figure C.1 is valid here, except for the use of a "mask element" to validate a desired number of specific positions within the "structuring element" or neighborhood. (It identifies the "care" positions and the "don't care" positions.) For the transformations, the .EXOR. operation is computed first, between the rows of the structuring element window and those of the image window. The result is then .AND.'ed with the rows of the mask window. Here too, the comparison with all rows of structuring element and mask windows for shifts 0, 1, and, 2, are computed and one row of output image is generated, before shifting the image window down the image one row. The processing continues until the last row of output image is completed. While

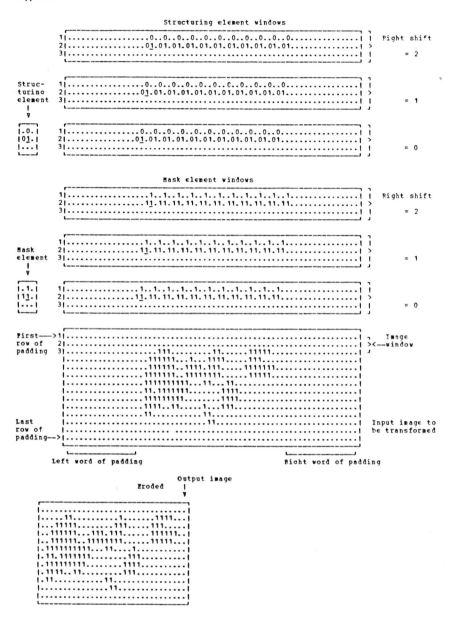

Figure C.2. Example of "parallel" processing of a binary compressed image in GIAPP.

.EXOR. detects the mismatches, the subsequent .AND. detects only the desired mismathes. For both erosion and dilatation, there must be no mismatch (i.e., there must be a perfect match) in this instance between the structuring element and the corresponding neighborhood in the image. Both the structuring element and the mask element are entered with values of 0 or 1 for each pixel. However, the "don't care" pixels or positions correspond to the 0 pixels in the mask.

Erosions and dilatations for hexagonal raster images (e.g., the transformations illustrated in Figure 3.5) are processed in a manner identical to what was described in Figures C.1 and C.2. However, two different structuring elements (different configurations of the same element) and mask elements are needed for processing even and odd rows. This is done to obtain the desired neighborhood configuration for the two cases. The hexagonal raster is exemplified in Figure 2.1

An Interactive
Session In Giapp

To show the interaction with the image-analysis programs available on a CYBER general-purpose computer, a sequence of operational steps is illustrated in Figure D.1.

In brief, the interactive steps are as follows:

I. Initial procedure for fetching and selecting the desired data (images) and program files.

II. Display of menu of options for processing.

III. Initialization procedure for building a directory of input/output files for the session.

IV. Display of the directory in an abbreviated form (nine files are available on input unit 1: the data set (DS) numbers are 100 to 108).

V. Display of binary image DS = 106. The image has a square raster.

VI. Redisplay of image in DS = 106, with a larger dot spacing between pixels on the monitor).

VII. Creation of an octagonal, 5 × 5, structuring element and its storage as an image in DS=200.

VIII. Computation of structuring-element window for the image to be transformed, which is 180 pixels wide. The value of 180 pixels is known beforehand in this instance. The window consists of 37 structuring elements side by side, and is 5 rows deep.

211

Figure D.1. The actual display on a Tektronix 4014/1 graphic terminal during a 20-step interactive execution. Note that each step is labeled by the corresponding Roman numeral. The small arrows identify the input commands keyed in by the user. A Tektronix 4631 hard-copy unit was used to obtain photocopies of the interaction from the terminal.

```
XXM01 GIAPP
TLLIO TYPE LLIO
IW: 1=ALL, 2=EXISTING FILES, 3=GIVEN DATA SET(KDS)
    4=GIVEN I IN LLIO(I, )   SET SW4 TO CUT, IW=NOT 1TO4:EXIT
IW:I12
LLIO
 1    9    0    0    0    0 0 0 0 0 0
 2    0    0    0    0    0 0 0 0 0 0          IV
 3    0    0    0    0    0 0 0 0 0 0
 4    0    0    0    0    0 0 0 0 0 0

 5  100 11 1 1  1  1    0    0 1 40
 6  101 11 1 1  1  2    0    0 1 40
 7  102 11 1 1  1  3    0    0 1 40
 8  103 11 1 1  1  4    0    0 1 40
 9  104 11 1 1  1  5    0    0 1 40
10  105 11 1 1  1  6    0    0 1 40
11  106 11 1 1  1  7    0    0 1 40
12  107 11 1 1  1  8    0    0 1 40
13  108 11 1 1  1  9    0    0 1 40
XXM00: G I A P PXX
MODE OF OPERATION?
0 - MANUAL OPERATION
1 - MENU DRIVEN, NO MENU DISPLAY
2 - DISPLAY MENU
>>> TYPE YOUR CHOICE.18
XXM02 GIAPP
PLOT COMPRESSED IMAGE ON TEKRONIX
DO YOU WANT TO SET PARAMETERS?
0=NO, 1=YES
ERASE SCREEN AND TYPE 1 TO CONTINUE.1               V
```

```
                    XXX PARAMETER  SETTINGS XXX

         TYPE OF SETTING                              CURRENT VALUE
                                                      -------------
  0   NO MORE SETTINGS
  1   DATA SET TO DISPLAY......................         300
  2   BAUD RATE (30 TO 960 CHARS/SEC).........         960
  3   BLACK ON WHITE - 1......................           1
      WHITE ON BLACK - 2
  4   SQUARE   PLOT - 1 OR 2..................           2
      HEXAGONAL PLOT - 3
  5   VECTOR DISPLAY(FAST) - 1................           1
      POINT  DISPLAY(SLOW) - 2
  6   SPACING BETWEEN PIXELS (USE A REAL VALUE)........  3.0
  7   MODEL OF TEKTRONIX......................           3
            1 - MOD 4006,4010 AND 4012/4013
            2 - MOD 4014/4015
            3 - MOD 4014/4015 + ENHANCED GRAPHICS MODE
  8   ADDRESSABLE POINTS......................           2
            1 - 1024 ADDRESSABLE POINTS
            2 - 4096 ADDRESSABLE POINTS
  9   SCREEN X-COORD. TOP LEFT CORNER.........         100
 10   SCREEN Y-COORD. TOP LEFT CORNER.........        3050
 11   REDISPLAY THIS MENU

>>> TYPE YOUR CHOICE.1
>>> TYPE THE NEW VALUE OF THE PARAMETER.106
>>> TYPE YOUR CHOICE.6
>>> TYPE THE NEW VALUE OF THE PARAMETER.6.
>>> TYPE YOUR CHOICE.9
>>> TYPE THE NEW VALUE OF THE PARAMETER.200
>>> TYPE YOUR CHOICE.11
```

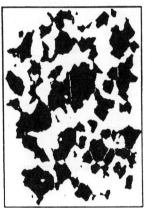

213

Figure D.1. (*continued*)

```
            ::: PARAMETER  SETTINGS :::

      TYPE OF SETTING                          CURRENT VALUE
                                               -------------
    0   NO MORE SETTINGS
    1   DATA SET TO DISPLAY.........................       106
    2   BAUD RATE (30 TO 960 CHARS/SEC)............       960
    3   BLACK ON WHITE = 1.........................         1
        WHITE ON BLACK = 2
    4   SQUARE    PLOT = 1 OR 2....................         2
        HEXAGONAL PLOT = 3
    5   VECTOR DISPLAY(FAST) = 1...................         1
        POINT  DISPLAY(SLOW) = 2
    6   SPACING BETWEEN PIXELS (USE A REAL VALUE)..       6.0
    7   MODEL OF TEKTRONIX.........................         3
        1 = MOD 4006,4010 AND 4012/4013
        2 = MOD 4014/4015
        3 = MOD 4014/4015 + ENHANCED GRAPHICS MODE
    8   ADDRESSABLE POINTS.........................         2
        1 = 1024 ADDRESSABLE POINTS
        2 = 4096 ADDRESSABLE POINTS
    9   SCREEN X-COORD. TOP LEFT CORNER...........       200
   10   SCREEN Y-COORD. TOP LEFT CORNER...........       3050
   11   REDISPLAY THIS MENU

>>> TYPE YOUR CHOICE.0
PICTC   106   200   3050     6.000000        1   2   1   5   0
              1  4095  3071   960     3
              2  4096
TYPE 1 TO CONTINUE.1
::M00: G I A P ::
  MODE OF OPERATION?
  0 = MANUAL OPERATION
  1 = MENU DRIVEN, NO MENU DISPLAY
  2 = DISPLAY MENU1
  >>> TYPE YOUR CHOICE.18
::M02 GIAPP
  PLOT COMPRESSED IMAGE ON TEKRONI1               VI
  DO YOU WANT TO SET PARAMETERS?
  0=NO, 1=YES1
ERASE SCREEN AND TYPE 1 TO CONTINUE.1
            ::: PARAMETER  SETTINGS :::

      TYPE OF SETTING                          CURRENT VALUE

      ---------------                          -------------
    0   NO MORE SETTINGS
    1   DATA SET TO DISPLAY.........................       106
    2   BAUD RATE (30 TO 960 CHARS/SEC)............       960
    3   BLACK ON WHITE = 1.........................         1
        WHITE ON BLACK = 2
    4   SQUARE    PLOT = 1 OR 2....................         2
        HEXAGONAL PLOT = 3
    5   VECTOR DISPLAY(FAST) = 1...................         1
        POINT  DISPLAY(SLOW) = 2
    6   SPACING BETWEEN PIXELS (USE A REAL VALUE)..       6.0
    7   MODEL OF TEKTRONIX.........................         3
        1 = MOD 4006,4010 AND 4012/4013
        2 = MOD 4014/4015
        3 = MOD 4014/4015 + ENHANCED GRAPHICS MODE
    8   ADDRESSABLE POINTS.........................         2
        1 = 1024 ADDRESSABLE POINTS
        2 = 4096 ADDRESSABLE POINTS
    9   SCREEN X-COORD. TOP LEFT CORNER...........       200
   10   SCREEN Y-COORD. TOP LEFT CORNER...........       3050
   11   REDISPLAY THIS MENU

>>> TYPE YOUR CHOICE.6
>>> TYPE THE NEW VALUE OF THE PARAMETER.9.
>>> TYPE YOUR CHOICE.0
PICTC   106   200   3050     9.000000        1   2   1
  5     2
              1  4095  3071   960     3
              2  4096
TYPE 1 TO CONTINUE.1
::M00: G I A P ::
  MODE OF OPERATION?
  0 = MANUAL OPERATION
  1 = MENU DRIVEN, NO MENU DISPLAY
  2 = DISPLAY MENU1
  >>> TYPE YOUR CHOICE.13            VII
::M02 GIAPP
  CREATE STRUCTURING ELEMENTS
  DO YOU WANT TO SET PARAMETERS?
  0=NO, 1=YES1
ERASE SCREEN AND TYPE 1 TO CONTINUE.1
            ::: PARAMETER SETTINGS :::

      TYPE OF SETTING                          CURRENT VALUE

      ---------------                          -------------
    0   NO MORE SETTINGS
    1   NUMBER OF COLUMNS IN 1 ROW OF PICTURE..............
  5
    2   NUMBER OF ROWS IN 1 PICTURE.......................
  5
    3   TYPE OF INPUT....................................
  1
        1 = STRUCTURING ELEMENT INPUT
        2 = MASK INPUT
    4   REDISPLAY THIS MENU

>>> TYPE YOUR CHOICE.0
HPR49    5    5    1
```

214

```
GIVE DS NO.:(I3)200
       200
IS IT OK& 1=YES, 2=NO, 3=ABORT|
GIVE NAME OF DATA:20 CHARACTERS octagon
>PICT=0005 0005      OCTAGON
IS IT OK& 1=YES, 2=NO, 3=ABORT|
PLEASE ENTER STRUCTURING ELEMENT:ISI1
01110
11111
11111
11111
01110
THE STRUCTURING ELEMENT IS:


01110
11111
11111
11111
01110

IS IT OK& 1=YES, 2=NO, 3=ABORT|
```

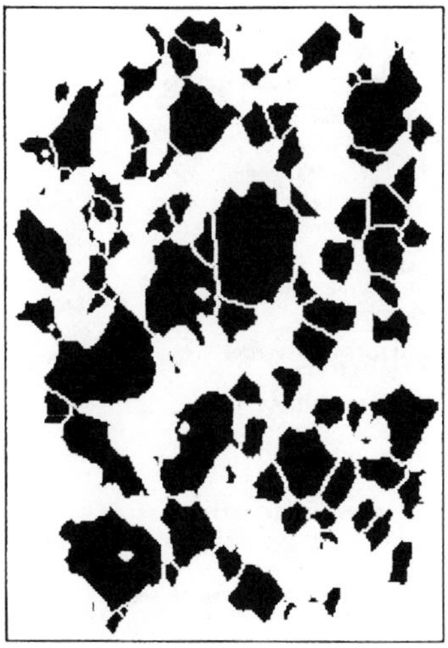

VII

```
>PICT=0005 0005      OCTAGON
```

```
$IM00: G I A P P$$
MODE OF OPERATION?
0 - MANUAL OPERATION
1 - MENU DRIVEN, NO MENU DISPLAY
2 - DISPLAY MENU|
>>> TYPE YOUR CHOICE.14
$IM02 GIAPP
```

VIII

```
CREATE STRUCTURING ELEMENT OR MASK WINDOW
DO YOU WANT TO SET PARAMETERS?
0=NO, 1=YES|
ERASE SCREEN AND TYPE 1 TO CONTINUE.|
                  $$$ PARAMETER  SETTINGS$$
```

TYPE OF SETTING	CURRENT VALUE
0 NO MORE SETTINGS	
1 INPUT DATA SET FOR STR ELEM OR MASK.................	200
2 TYPE OF RASTER..	2
1,2 = SQUARE	
3 = HEXAGONAL	
3 BITS/ROW OF THE BINARY IMAGE......................	0
4 WHAT TO LOAD...	1
1 = STRUCTURING ELEMENT WINDOW	
2 = MASK WINDOW	
5 COLUMNS IN STRUCTURING ELEMENT OR MASK..............	3
6 ROWS IN STRUCTURING ELEMENT OR MASK.................	3
7 WORDS IN ONE ROW OF STRUCTURING ELEMENT OR MASK....	20
8 REDISPLAY THIS MENU	

```
>>> TYPE YOUR CHOICE.1
>>> TYPE THE NEW VALUE OF THE PARAMETER.200
>>> TYPE YOUR CHOICE.3
>>> TYPE THE NEW VALUE OF THE PARAMETER.180
>>> TYPE YOUR CHOICE.5
>>> TYPE THE NEW VALUE OF THE PARAMETER.5
>>> TYPE YOUR CHOICE.6
>>> TYPE THE NEW VALUE OF THE PARAMETER.5
>>> TYPE YOUR CHOICE.0
=PR92    200       2     180      1      5      5     20
S E STRING:NO. OF BITS=    300    NO. OF ELEMENTS=    37
IDIMBS=    59   IDIMBE=    243
NUMBER OF 1 BITS IN S E SET =        777.0
$IM00: G I A P P$$
MODE OF OPERATION?
0 - MANUAL OPERATION
1 - MENU DRIVEN, NO MENU DISPLAY
2 - DISPLAY MENU|
>>> TYPE YOUR CHOICE.15
```

215

Figure D.1. (continued)

```
±±M02 GIAPP
MINKOWSKI TRANSFORMATION (ONLY BLACK)
DO YOU WANT TO SET PARAMETERS?
0=NO, 1=YES1                                                        IX
ERASE SCREEN AND TYPE 1 TO CONTINUE.1
                              ±±± PARAMETER SETTINGS ±±±

                TYPE OF SETTING                            CURRENT VALUE
                ---------------                            -------------
           0    NO MORE SETTINGS
           1    INPUT DATA SET FOR IMAGE TO BE TRANSFORMED........  200
           2    INPUT DATA SET FOR STRUCTURING ELEMENT...........   200
           3    OUTPUT DATA SET..................................   106
           4    TYPE OF TRANSFORMATION...........................     1
                   1 = DILATATION
                   2 = EROSION
           5    TYPE OF RASTER...................................     2
                      1,2 = SQUARE
                      3   = HEXAGONAL
           6    REDISPLAY THIS MENU

    >>> TYPE YOUR CHOICE.1
    >>> TYPE THE NEW VALUE OF THE PARAMETER.106
    >>> TYPE YOUR CHOICE.3
    >>> TYPE THE NEW VALUE OF THE PARAMETER.300
    >>> TYPE YOUR CHOICE.4
    >>> TYPE THE NEW VALUE OF THE PARAMETER.2
    >>> TYPE YOUR CHOICE.0
MINK1    106    200    300    2     2    180    5
          5    300    59   243   20    5
KMASK1:RANGE    59   243
TOTAL BITS IN IUNA, AND IN JUN
    =    17344.        8996.
±±M00: G  I  A  P P±±
MODE OF OPERATION?
0 = MANUAL OPERATION
1 = MENU DRIVEN, NO MENU DISPLAY
2 = DISPLAY MENU1
    >>> TYPE YOUR CHOICE.15
±±M02 GIAPP
MINKOWSKI TRANSFORMATION (ONLY BLACK)
DO YOU WANT TO SET PARAMETERS?
0=NO, 1=YES1
ERASE SCREEN AND TYPE 1 TO CONTINUE.1


                        ±±± PARAMETER SETTINGS ±±±              X

                TYPE OF SETTING                            CURRENT VALUE
                ---------------                            -------------
           0    NO MORE SETTINGS
           1    INPUT DATA SET FOR IMAGE TO BE TRANSFORMED........  106
           2    INPUT DATA SET FOR STRUCTURING ELEMENT...........   200
           3    OUTPUT DATA SET..................................   300
           4    TYPE OF TRANSFORMATION...........................     2
                   1 = DILATATION
                   2 = EROSION
           5    TYPE OF RASTER...................................     2
                      1,2 = SQUARE
                      3   = HEXAGONAL
           6    REDISPLAY THIS MENU

    >>> TYPE YOUR CHOICE.1
    >>> TYPE THE NEW VALUE OF THE PARAMETER.300
    >>> TYPE YOUR CHOICE.3
    >>> TYPE THE NEW VALUE OF THE PARAMETER.400
    >>> TYPE YOUR CHOICE.4
    >>> TYPE THE NEW VALUE OF THE PARAMETER.1
    >>> TYPE YOUR CHOICE.0
MINK1    300    200    400    1     2    180    5
          5    300    59   243   20    5
KMASK1:RANGE    59   243
TOTAL BITS IN IUNA, AND IN JUN
    =    8996.        15935.
±±M00: G  I  A  P P±±
MODE OF OPERATION?
0 = MANUAL OPERATION
1 = MENU DRIVEN, NO MENU DISPLAY
2 = DISPLAY MENU1
    >>> TYPE YOUR CHOICE.18
±±M02 GIAPP
PLOT COMPRESSED IMAGE ON TEKRONIX
DO YOU WANT TO SET PARAMETERS?
0=NO, 1=YES0
PICTC    400    200    3050        9.000000       1   2   1   5   5
          1    4095   3071    960    3
          2    4096                                     XI
TYPE 1 TO CONTINUE.1
```

216

```
**M00: G I A P P**
 MODE OF OPERATION?
 0 = MANUAL OPERATION
 1 = MENU DRIVEN, NO MENU DISPLAY
 2 = DISPLAY MENU1
 >>> TYPE YOUR CHOICE.18
**M02 GIAPP
 PLOT COMPRESSED IMAGE ON TEKRONIX
 DO YOU WANT TO SET PARAMETERS?
 0=NO, 1=YES1
 ERASE SCREEN AND TYPE 1 TO CONTINUE.1
                    *** PARAMETER SETTINGS ***
```

XII

```
           TYPE OF SETTING                              CURRENT VALUE
           ---------------                              -------------

        0  NO MORE SETTINGS
        1  DATA SET TO DISPLAY................................     400
        2  BAUD RATE (30 TO 960 CHARS/SEC)...................     960
        3  BLACK ON WHITE = 1................................       1
           WHITE ON BLACK = 2
        4  SQUARE      PLOT = 1 OR 2.........................       2
           HEXAGONAL PLOT = 3
        5  VECTOR DISPLAY(FAST) = 1..........................       1
           POINT  DISPLAY(SLOW) = 2
        6  SPACING BETWEEN PIXELS (USE A REAL VALUE).........     9.0
        7  MODEL OF TEKTRONIX................................       3
               1 = MOD 4006,4010 AND 4012/4013
               2 = MOD 4014/4015
               3 = MOD 4014/4015 + ENHANCED GRAPHICS MODE
        8  ADDRESSABLE POINTS................................       2
               1 = 1024 ADDRESSABLE POINTS
               2 = 4096 ADDRESSABLE POINTS
        9  SCREEN X-COORD. TOP LEFT CORNER...................     200
       10  SCREEN Y-COORD. TOP LEFT CORNER...................    3050
       11  REDISPLAY THIS MENU

   >>> TYPE YOUR CHOICE.1
   >>> TYPE THE NEW VALUE OF THE PARAMETER.300
   >>> TYPE YOUR CHOICE.0
   PICTC   300   200   3050      9.000000        1    2    1
   5   6
               1  4095  3071   960   3
               2  4096
   TYPE 1 TO CONTINUE.1
```

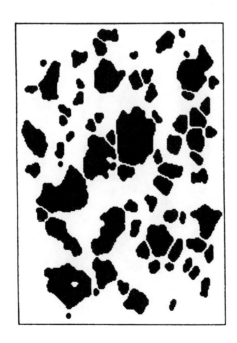

Figure D.1. *(continued)*

```
**M00: G I A P P**
MODE OF OPERATION?
0 = MANUAL OPERATION
1 = MENU DRIVEN, NO MENU DISPLAY
2 = DISPLAY MENU1
>>> TYPE YOUR CHOICE.12
**M02 GIAPP
LOGICAL OPERATIONS ON COMPRESSED IMAGES
DO YOU WANT TO SET PARAMETERS?
0=NO, 1=YES1
ERASE SCREEN AND TYPE 1 TO CONTINUE.1
                      *** PARAMETER  SETTINGS***
```

XIII

```
        TYPE OF SETTING                              CURRENT VALUE

        ---------------                              -------------

        0  NO MORE SETTINGS
        1  FIRST INPUT PICTURE DS. NUMBER...................
  1
        2  SECOND INPUT PICTURE DS. NUMBER...................    2
 00
        3  OUTPUT PICTURE DS. NUMBER.........................    3
 00
        4  TYPE OF OPERATION................................
  1
                     1 = A.B
                     2 = A+B
                     3 = A(+)B
                     4 = NOT(A)
                     5 = NOT(A.B)
                     6 = NOT(A+B)
                     7 = NOT(A(+)B
                     8 = A.NOT(B)
                     9 = A+NOT(B)
                    10 = A(+)NOT(B)
                  5  REDISPLAY THIS MENU

>>> TYPE YOUR CHOICE.1
>>> TYPE THE NEW VALUE OF THE PARAMETER.106
>>> TYPE YOUR CHOICE.2
>>> TYPE THE NEW VALUE OF THE PARAMETER.400
>>> TYPE YOUR CHOICE.3
>>> TYPE THE NEW VALUE OF THE PARAMETER.500
>>> TYPE YOUR CHOICE.4
>>> TYPE THE NEW VALUE OF THE PARAMETER.8
>>> TYPE YOUR CHOICE.0
*LOGO3   106    400    500      8
KMASK    180    18
TOTAL BITS IN IUNA,IUNB,JUN=
         17344.0        15935.0        1409.0
```

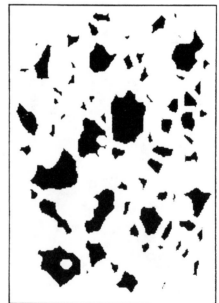

```
**M00: G I A P P**
MODE OF OPERATION?
0 = MANUAL OPERATION
1 = MENU DRIVEN, NO MENU DISPLAY
2 = DISPLAY MENU
>>> TYPE YOUR CHOICE.18
**M02 GIAPP
PLOT COMPRESSED IMAGE ON TEKRONIX
DO YOU WANT TO SET PARAMETERS?
0=NO, 1=YES0
```

XIV

```
**M00: G I A P P**
MODE OF OPERATION?
0 = MANUAL OPERATION
1 = MENU DRIVEN, NO MENU DISPLAY
2 = DISPLAY MENU
>>> TYPE YOUR CHOICE.2
**M01 GIAPP
COPY COMPRESSED DATA:GIVE IUN,JUN:213300201
*COPY1  300  201
```

XV

```
**M00: G I A P P**
MODE OF OPERATION?
0 = MANUAL OPERATION
1 = MENU DRIVEN, NO MENU DISPLAY
2 = DISPLAY MENU
>>> TYPE YOUR CHOICE.2
**M01 GIAPP
COPY COMPRESSED DATA:GIVE IUN,JUN:213400202
*COPY1  400  202
**M00: G I A P P**
MODE OF OPERATION?
0 = MANUAL OPERATION
1 = MENU DRIVEN, NO MENU DISPLAY
2 = DISPLAY MENU
>>> TYPE YOUR CHOICE.2
**M01 GIAPP
COPY COMPRESSED DATA:GIVE IUN,JUN:213500203
*COPY1  500  203
**M00: G I A P P**
MODE OF OPERATION?
0 = MANUAL OPERATION
1 = MENU DRIVEN, NO MENU DISPLAY
2 = DISPLAY MENU
>>> TYPE YOUR CHOICE.4
**M01 GIAPP
RUB:GIVE IUN:13300
RUB  300
**M00: G I A P P**
MODE OF OPERATION?
0 = MANUAL OPERATION
1 = MENU DRIVEN, NO MENU DISPLAY
2 = DISPLAY MENU
>>> TYPE YOUR CHOICE.2
**M01 GIAPP
COPY COMPRESSED DATA:GIVE IUN,JUN:213400300
*COPY1  400  300
**M00: G I A P P**
MODE OF OPERATION?
0 = MANUAL OPERATION
1 = MENU DRIVEN, NO MENU DISPLAY
2 = DISPLAY MENU
>>> TYPE YOUR CHOICE.10
**M01 GIAPP
CORRELATION BETWEEN BINARY IMAGES
DO YOU WANT TO SET PARAMETERS?
0=NO, 1=YES1
```

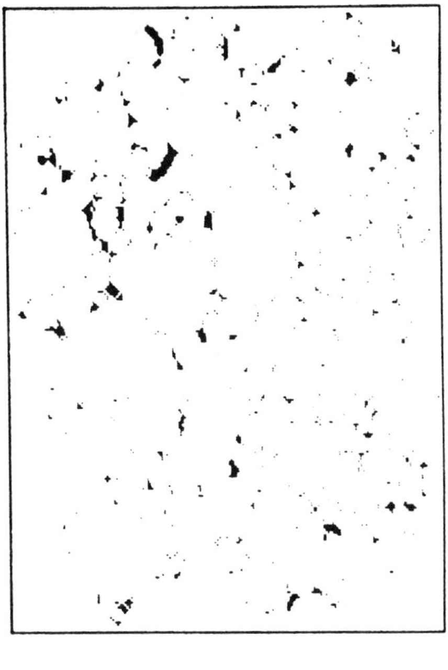

Figure D.1. *(continued)*

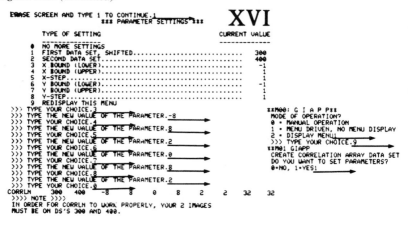

```
ERASE SCREEN AND TYPE 1 TO CONTINUE.1
                    ### PARAMETER SETTINGS ###          XVI

        TYPE OF SETTING                            CURRENT VALUE
        ---------------                            -------------
    0   NO MORE SETTINGS
    1   FIRST DATA SET, SHIFTED............................  300
    2   SECOND DATA SET...................................  400
    3   X BOUND (LOWER)...................................   -1
    4   X BOUND (UPPER)...................................    1
    5   X-STEP............................................    1
    6   Y BOUND (LOWER)...................................   -1
    7   Y BOUND (UPPER)...................................    1
    8   Y-STEP............................................    1
    9   REDISPLAY THIS MENU
>>> TYPE YOUR CHOICE.3                          ##M00: G I A P P##
>>> TYPE THE NEW VALUE OF THE PARAMETER.-8        MODE OF OPERATION?
>>> TYPE YOUR CHOICE.4                            0 - MANUAL OPERATION
>>> TYPE THE NEW VALUE OF THE PARAMETER.8         1 - MENU DRIVEN, NO MENU DISPLAY
>>> TYPE YOUR CHOICE.5                            2 - DISPLAY MENU
>>> TYPE THE NEW VALUE OF THE PARAMETER.2         >>> TYPE YOUR CHOICE.9
>>> TYPE YOUR CHOICE.6                          ##M01 GIAPP
>>> TYPE THE NEW VALUE OF THE PARAMETER.0         CREATE CORRELATION ARRAY DATA SET
>>> TYPE YOUR CHOICE.7                            DO YOU WANT TO SET PARAMETERS?
>>> TYPE THE NEW VALUE OF THE PARAMETER.8         0-NO, 1-YES1
>>> TYPE YOUR CHOICE.8
>>> TYPE THE NEW VALUE OF THE PARAMETER.2
>>> TYPE YOUR CHOICE.0
CORRLN   300   400   -8    8    0    8    2    32    32
>>>> NOTE >>>>
IN ORDER FOR CORRLN TO WORK PROPERLY, YOUR 2 IMAGES
MUST BE ON DS'S 300 AND 400.
```

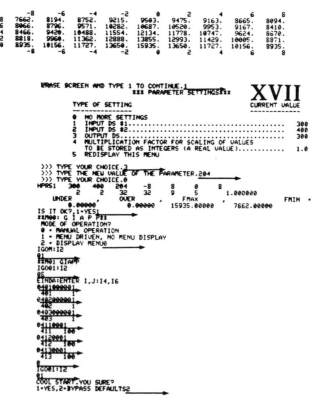

```
        -8      -6      -4      -2      0       2       4       6       8
  8    7662.   8194.   8752.   9215.   9503.   9475.   9163.   8665.   8094.
  6    8066.   8796.   9571.  10282.  10687.  10520.   9953.   9167.   8410.
  4    8466.   9420.  10488.  11554.  12134.  11778.  10747.   9624.   8670.
  2    8818.   9960.  11362.  12888.  13855.  12993.  11429.  10005.   8871.
  0    8935.  10156.  11727.  13650.  15935.  13650.  11727.  10156.   8935.
        -8      -6      -4      -2      0       2       4       6       8

               ERASE SCREEN AND TYPE 1 TO CONTINUE.1
                        ### PARAMETER SETTINGS###          XVII

            TYPE OF SETTING                            CURRENT VALUE
            ---------------                            -------------
        0   NO MORE SETTINGS
        1   INPUT DS #1.......................................  300
        2   INPUT DS #2.......................................  400
        3   OUTPUT DS.........................................  300
        4   MULTIPLICATION FACTOR FOR SCALING OF VALUES
            TO BE STORED AS INTEGERS (A REAL VALUE)............  1.0
        5   REDISPLAY THIS MENU

        >>> TYPE YOUR CHOICE.3
        >>> TYPE THE NEW VALUE OF THE PARAMETER.204
        >>> TYPE YOUR CHOICE.0
        HPR51   300   400   204   -8    8    0    8
                  2    2    32    32    9    5    1.000000
            UNDER           OVER           FMAX          1.000000
        0.00000         0.00000        15935.00000    7662.00000      FMIN =
        IS IT OK?,1-YES1
        ##M00: G I A P P##
        MODE OF OPERATION?
        0 - MANUAL OPERATION
        1 - MENU DRIVEN, NO MENU DISPLAY
        2 - DISPLAY MENU0
        IGOM:I2
        01
        ##M01 GIAPP
        IGO01:I2
        05
        EINDATENTER I,J:I4,I6
        0401000001
           401    1
        0402000001
           402    1
        0403000001
           403    1
        0411000001
           411   100
        0412000001
           412   100
        0413000001
           413   100
        0
        IGO01:I2
        01
        COOL START.YOU SURE?
        1-YES,2-BYPASS DEFAULTS2
```

```
ELLIO  30  14   2   2   3
PREPD   0   0  13  14  17
RLIO1, RECON LLIO FROM MT- 11          XVIII

                         .

RLIO1, RECON LLIO FROM MT- 11
DATA START
-----
FILE,DS-   1 100
       1   100   12   40   1    0    0    0    0   180
      18   252
-1  5 >PICT-0180 0252  ID-CALCIC PYROXENE
-----
FILE,DS-   2 101
       2   101   12   40   1    0    0    0    0   180
      18   252
-1  5 >PICT-0180 0252  ID-CRYSTAL2
-----
FILE,DS-   3 102
       3   102   12   40   1    0    0    0    0   180
      18   252
-1  5 >PICT-0180 0252  ID-CRYSTAL3
-----
FILE,DS-   4 103
       4   103   12   40   1    0    0    0    0   180
      18   252
-1  5 >PICT-0180 0252  ID-CRYSTAL4
-----
FILE,DS-   5 104
       5   104   12   40   1    0    0    0    0   180
      18   252
-1  5 >PICT-0180 0252  ID-CRYSTAL5
-----
FILE,DS-   6 105
       6   105   12   40   1    0    0    0    0   180
      18   252
-1  5 >PICT-0180 0252  ID-CRYSTAL6
-----
FILE,DS-   7 106
       7   106   12   40   1    0    0    0    0   180
      18   252
-1  5 >PICT-0180 0252  ID-CRYSTAL7
-----
FILE,DS-   8 107
       8   107   12   40   1    0    0    0    0   180
      18   252
-1  5 >PICT-0180 0252  ID-CRYSTAL8
-----
FILE,DS-   9 108
       9   108   12   40   1    0    0    0    0   180
      18   252
-1  5 >PICT-0180 0252  ID-CRYSTAL9
DATA END
FILE,DS-   5 204
       5   204   29   10   1    0    0    0    0    0
       9    5  300  400  -8    8    0    8    2    2
    1000   18  252  180   40   18  252  180   40
  0  5 >PICT-0180 0252   ID-CRYSTAL7
  1  5   MINK1-     EROSION
  2  5 >PICT-0005 0005       OCTAGON
  3  5   MINK1-  DILATATION
  4  5 >PICT-0005 0005       OCTAGON
  5  5    ****


  0  5 >PICT-0180 0252  ID-CRYSTAL7
  1  5   MINK1-     EROSION
  2  5 >PICT-0005 0005       OCTAGON
  3  5   MINK1-  DILATATION
  4  5 >PICT-0005 0005       OCTAGON
 -1  5   CORRLN


DATA END
```

Figure D.1. *(continued)*

ₐLIO1, RECON LLIO FROM MT- 12

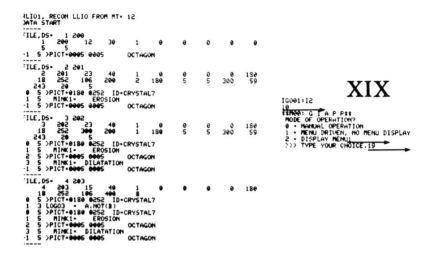

IX. Computation of an erosion for the image in DS = 106, by the octagonal structuring element stored in DS = 200. It produces a transformed (eroded) image stored in DS = 300. There are 17,344 binary 1 pixels in the original image, and 8996 binary 1 pixels in the eroded image.

X. Dilatation of the eroded image in DS = 300 by the octagonal structuring element in DS = 200. The result, an "opened" image, is stored in DS = 400. The number of binary 1 pixels in the latter opened image is 13,935 (there are 17,344 binary 1 pixels in the original image).

XI. Display of the closed image in DS = 400.

XII. Display of the eroded image in DS = 300.

XIII. Logical operation (A).AND..NOT.(B) between the original image in DS = 106 (A) and the closed image in DS = 400 (B). The operation produces a new image in DS = 500 with the "opening residue." There are 1409 binary 1 pixels in this image.

XIV. Display of the image of the closing "residue" in DS = 500.

XV. Saving (copying) of data sets 300, 400, and 500 into DSs 201, 202, and 203, respectively, elimination of DS = 300, and copying of DS = 400 (the opened image) into DS = 300. Now, two copies of the opened image reside in DSs 300 and 400.

XVI. Computation of the autocorrelation (or geometric autocovariance) for the closed image in DS = 300 and DS = 400, for shifts of two pixels between image coordinates −8 and 8 (horizontally) and between 0 and 8 (vertically). The value at (0,0), for the unshifted coincident position for the two images, is 15,935, that is, the number of binary 1 pixels in the closed image.

XVII. Storage of the correlation array as a gray-level image in DS = 204.

XVIII. Entering "manual" mode of interaction, input of parameters for activating printing of the complete image information, which consists of "header" and "commentaries" for all the available data sets in the 100s and 200s series.

XIX. Return to "menu-driven" interaction and exit from GIAPP.

XX. Permanent disk storage of the collected results (images in DSs 200s) now on an output unit termed "tape 12." The saved file (catalogued) is termed "session 1, id=mt12." It can be reused in other sessions and more image files can be added to it.

REFERENCES

Fabbri, A. G., 1980, GIAPP: Geological Image Analysis Program Package for Estimating Geometrical Probabilities, *Comput. Geosci.* **6:**153-161.

Fabbri, A. G., 1981, Image Processing of Coincident Binary Patterns from Geological and Geophysical Maps of Mineralized Areas, in *Proc. Canadian Man-Computer Comm. Soc., 7th Conf., June 10-12, 1981, Waterloo, Ontario,* pp. 323-331; also in *Uranium in Granites,* Y. T. Maurice, ed., 1982, Geol. Surv. Can., Paper 81-23, pp. 157-165.

Fabbri, A. G., T. Kasvand, and J. H. Stray, 1978, Implementation of an Interactive System for Computer Processing of Geological Images, in Current Research, Part C., *Geol. Surv. Can. Paper 78-1C,* pp. 123-124.

Rosenfeld, A., and A. C. Kak, 1976, *Digital Picture Processing,* Academic Press, New York, 457p.

GLOSSARY

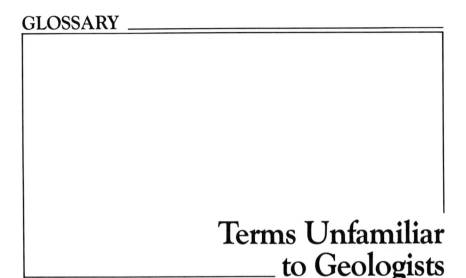

Terms Unfamiliar
to Geologists

Adjacency. A topological property of digital image subsets—pixels—which are termed neighbors: e. g., four neighbors (up, down, left, and right) and eight neighbors (all the external pixels in a 3 X 3 array) in a square raster image, or six neighbors in a hexagonal raster image (Rosenfeld, A., and A. C. Kak, 1976, *Digital Picture Processing*, Academic Press, New York, pp. 335-336.).

Algorithm. A precise sequence of steps that defines a specific computation; a general method of solution to a computable problem.

Analog techniques. Techniques that use analog computers, which operate by translating numbers into measurable quantities, such as voltages, resistency, rotations, or vice versa. An analog computer measures continuously, whereas a digital computer counts discretely.

.AND. A logical operator that computes a word as the result of the intersection between two words.

Annealing. Heating and holding at suitable temperature and then cooling at a suitable rate for complete recrystallization.

Artificial intelligence. Research and study in methods for the development of a machine that can improve its own operations; also the part of computer science concerned with designing intelligent computer systems.

Assembly language. A language similar in structure to a machine language, but made up of mnemonics and symbols. It represents a direct symbolic representation of the binary instruction that can be executed by the processor.

Automatic classification. The target of pattern recognition: the identification performed by a machine.

Batch mode. Processing in a sequential procedure that involves an accumulation or group of units, in contrast to online processing.

Baud (rate). The rate of sending signals down a communication line. One baud means one signal per second. For telephone lines, baud rate is convertible to characters per second by dividing by ten. For example, 300 bauds works out to about 30 characters per second.

Because a signal is any potential change from one frequency to another, during transmission each bit is converted to one or to the other frequency in use, so that each bit becomes a signal.

Binary. A number system with two digits, 0 and 1. Each distinct binary number represents a power of 2.

Binary compressed image. A binary image in which one bit represents the value stored for each pixel.

Binary code. *See* Machine-dependent language.

Binary expanded image. A binary image in which one computer word represents the value stored for each pixel.

Binary image. *See* Digital images.

Bit. Binary digit, a 0 or a 1: the smallest amount of information that a computer can hold. Bits can be grouped to form larger values, such as nibbles, bytes or words, in which numbers and characters can be stored.

Bit shift operator. A function that shifts the bits within one (or more) computer word to the left or to the right, one bit at a time.

Bit status. The value of a bit: 0 or 1 (OFF or ON).

Boolean algebra. A value that may be true or false; also referred to as logical algebra.

Boolean operations. *See* Logical operation.

Boolean operators. *See* Logical operator.

bpi. *Bits per inch:* density of the information on magnetic tape. A common density is 1600 bpi.

CA. *See* Cellular automaton.

Card reader. A device that inputs information stored on cards.

"Care" positions. Within a neighborhood representing a structuring element, the pixels being matched or mismatched with the corresponding pixels in the image being processed. *See also* "Don't care" positions.

Cathode ray tube. *See* CRT.

Cellular automaton (CA). An array of computers on which cellular logic is implemented.

Cellular logic. An operation performed digitally on an array of data $P(I,J)$ which is carried out so as to transform $P(I,J)$ into a new array $P'(I,J)$ wherein each element in the new array has a value determined only by the corresponding element in the original array along with the values of its nearest neighbors.

Center bit (or pixel). The bit (pixel) used to map the neighborhood of bits (pixels) that represent a structuring element.

Character recognition. The automatic recognition of printed and handwritten characters.

Characters per second. *See* Bauds.

C.I. *See* Complexity index.

Closing. A dilatation and then an erosion performed on a set (image) by a structuring element. First the image is dilatated and then the dilatated image is eroded.

Commentary. The set of comment records that are part of an image data set.

(DS) in GIAPP. It describes the original image name and the history of processing: each program writes its name and action in the commentary of the image processed.

Comment record. One row of information in the commentary of an image data set.

Complement. Logical negation: the complement of a set A, denoted A^c, is the set of points that are not in A.

Complexity index (C.I.). The ratio between the total cirfumference and the total area of the grain profiles for a given phase. (Underwood, E. E., 1970, *Quantitative Stereology,* Addison-Wesley, Reading, Mass., pp. 228-229).

Component. Image subset of connected pixels according to a given adjacency (four-neighbor, six-neighbor, or eight-neighbor adjacency). Also called *connected component.*

Component-labeled image. Image in which all components are identified by sequence numbers, or labels, which are the values of all the pixels belonging to each component.

Component labeling. The process of numbering the components in an image, so that for any component C (of a given image subset) all pixels of C have the same value, and no pixel not in C has that value.

Computer graphics. The field concerned primarily with the computer synthesis and manipulation of images specified by descriptions. Fundamental parts of the descriptions are coded as edges, polygons, and pointers to these within a specially designed data structure.

Computer word. *See* Word.

Connectivity number. The number of objects minus the number of holes in a binary image.

COOL START. An initialization procedure in GIAPP by which valid parameters are set (or reset) for properly running most programs during interaction. During the procedure a file occupancy table is constructed of all the data sets existing on the devices not used for scratch.

CRT. Cathode ray tube: the video display (television-like) part of a computer terminal.

Custer and complement operation. Transformation that detects successive edges of black objects in a binary image by erosions and complementations.

Custer operation. Transformation that detects the edges of black objects in a binary image by erosion.

DA. Digital-to-analog converter: a system that performs fast, real-time data conversion between digital and analog computers.

Data set. *See* DS.

Density. *See* Grey level.

Digital-to-analog converter. *See* DA.

Digital computer. A computer that operates by using numbers to express all the quantities and variables of a problem. The information processed is represented by combinations of discrete or discontinuous data as compared with an analog computer for continuous data.

Digital images. Rectangular arrays of pixels, generally integer arrays. Also digitized images.

Digitization. Sampling process used to extract from a picture a discrete set of real numbers (samples). *See also* Quantization.

Digitized images. *See* Digital images.

Digitizer. An instrument that performs digitization, e.g., a scanner, graphic tablet, or television camera.

Dilatation. A structuring-element transformation that changes the values of pixels in an image from 0 to 1 according to the local degree of matching between the pixels in the structuring element and the corresponding pixels in the image. The structuring element is translated so that its center pixel coincides in position with successively all the pixels in the image.

"Don't care" positions. Within a neighborhood representing a structuring element, the pixels not being matched nor mismatched with the corresponding pixels of the image being transformed.

DS. Data set of a computer file in GIAPP, in which image data is stored. The array of pixel values is preceded by header information and commentary data.

Eight-neighbor rule. A type of neighborhood transformation in which all eight immediately adjacent neighbors in a 3×3 neighborhood are considered.

Eight-neighbor shrinking. An erosion in which the structuring element consists of 3×3 black pixels.

Embedded Markov chain. Chain that does not recognize grain contacts among like species (phases).

Enclosed area labeling. *See* Component labeling.

Erosion. A structuring element transformation that changes the values of pixels in an image from 1 to 0 according to the local degree of matching between the pixels in the structuring element and the corresponding pixels in the image. The structuring element is translated so that its center pixel coincides in location with successively all the pixels in the image.

Exclusive .OR. *See* .EXOR.

.EXOR. A logical operator that computes a computer word as the result of the union of two non-overlapping subsets (bits) of two words.

Expansion. Computation that produces a binary expanded image from a binary compressed image.

Frame. Term used to indicate the pixels between the edges of the image and those pixels identifying the objects in the image.

Frame buffer. *See* Image buffer.

GIAPP. Geological Image Analysis Program Package for estimating geometrical probabilities.

Grain-transition probability. The probability of encountering a given grain when proceding from another grain.

Graphic terminal. Terminal hardware that enables the computer to display drawings and images.

Gray-level image. Black-and-white image in which there can be shades of gray but not color. The values of the pixels are called gray levels or brightness or gray tones of the image at the points corresponding to the pixels.

Gray-level slicing. *See* Thresholding.

Gray-level value. Intensity: value of a pixel in a gray-level image.

Hardware. Physical components and parts of a computer system.

Hard copy. Printout produced on paper, or any tangible permanent medium.

Header. Information preceding the pixel values which indicate how many pixels are stored in each row of image data (logical record) and how many rows are stored in the image file. It also indicates the type of image (binary, compressed, binary expanded, gray-level, etc.) and other information necessary to process the image.

Hexagonal raster. Hexagonal array of pixels.

Hexagonal scanning. Optical digitization according to a hexagonal array configuration.

High-level language. Language used to program computers that incorporates English-like statements and mathematical notation. FORTRAN and BASIC are examples of high-level languages.

High-pass filtering. Filtering in which high-pass spatial frequency is retained while low frequencies are suppressed.

Hit-or-miss transformation. A point-by-point transformation of sets or structuring elements transformation defined in mathematical morphology (Serra, J., 1982, *Image Analysis and Mathematical Morphology*, Academic Press, New York, pp. 34-62).

Image. *See* Digital images.

Image buffer. A large memory in which one or more digital images can be stored. Also called *frame buffer.*

Image processing. Operations that transform images into other images to obtain information about them and about the objects in them; operations that transform images into descriptions.

Image recognition. The mapping of images into image description.

Implicit storage of images. Method of storing images synthesized into a particular description understood by the computer program. The information is coded and sparse and is in the form of edges, polygons, and pointers.

228

Initialization procedure. The process of assigning initial values to variables. *See* COOL START.

Input/output (I/O). Input is data entered into the computer through a peripheral device, such as a card reader or a terminal; output is data transmitted by a computer to a peripheral device such as a line printer or a terminal. I/O refers to both types of data.

Integrated circuit (i.c.) Electronic circuit consisting of silicon chips. A single i.c. can contain from ten to ten thousand discrete electronic components.

Interactive. Conversational: the case in which the computer responds when prompted by a proper request.

Intercept. A measure of particle profile elongation in a particular direction.

Intersection. For any image subsets A and B, the intersection $A \cap B$ is the set of points that are in both A and B with the same value.

I/O. *See* Input/output.

Junction. The point at which segments of thinned boundary meet. A junction occurs whenever three profiles of grains are in contact with one another.

Label. The symbolic value assigned to a pixel by component labeling or by phase-labeling processes.

Light pen probe. Input device consisting of a pen that emits a light beam sensed by a TV monitor terminal. This probe enables the operator to interact with the computer for editing or measuring part of the image being displayed.

Local operators. *See* Neighborhood operator.

Logical operation. Operation between two computer words (or two groups of computer words) considered as sets of bits. Typical operations are union, intersection, complement, and combinations of these.

Logical operator. A single computer instruction of program-language statement for computing logical operations. Common logical operators are .OR., .AND., .EXOR., and .NOT.

Machine-dependent language. A low-level programming language, specific to each computer and consisting of 0's and 1's; also called *binary code.*

Markov chain. A chain of events in time or space where each event depends to a certain degree on a preceding event.

Markovian property. *See* Markovity.

Markovity. The similarity of grain sequences along linear traverses to Markov chains.

Mask element. A set of pixels within a neighborhood that are used to identify the "care" positions from the "don't care" positions in a structuring element for transformations in which both matching 0's and 1's are considered. *See also* Structuring element.

Match. Coincidence of value of pixels that are compared to each other.

Matching. A technique for finding the pixels where some given pattern (template, structuring element, a known object, or a piece of an image) occurs in the image.

Mathematical morphology. The study of geometric aspects of two-dimensional sets: granulometric properties and structural properties initiated by G. Matheron and J. Serra in Fountainebleau, France, that has provided a theoretical background to probabilistic concepts and applications to morphological measurements by image analysis.

Mean traverse. A measure of average elongation of a profile in a given direction obtained by averaging linear intercepts or chords across the profile.

Memory-type cathode ray tube. A CRT device that stores its display on the surface of the tube; opposite to a refreshed type of display device.

Menu-driven interaction. A type of "user-friendly" interaction in which a list of possible alternative operations or processing steps can be selected during a processing session.

Minkowski difference. *See* Erosion.

Minkowski sum. *See* Dilatation.

Minkowski transformation (of sets). *See* Minkowski-type operation.

Minkowski-type operation. Operation between sets that generate (and are the mathematical foundation of) the structuring-element transformations developed in mathematical morphology for characterizing geometric properties by probabilistic statements.

Monitor. A closed-circuit television receiver.

Morphology. Systematic study of form in general, and of its origin and physical relevance (of rocks, clouds, colors, plants, animals, etc.). In pattern recognition and image processing, it is the study of the visible makeup of a particular object or kind of object—that is, its shape—to classify or identify a characteristic form or structure.

Multiuser environment. A time-shared computer.

Negation. *See* Complement.

Neighbor. A pixel that is adjacent to another pixel. *See* Adjacency.

Neighborhood. A set of adjacent pixels.

Neighborhood operator. A program or single instruction that evaluated a function that for each pixel in an image, takes into account the values of a subset of neighboring pixels (not excluding the pixel itself). Such operators are often used to detect geometrical features to characterize an image or parts of it.

Nonbinary image. A gray-level image or a labeled image.

Non-overlapping subset. The result of an exclusive .OR.'ing logical operation between two computer words. *See also* .EXOR.

.NOT. The operator that computes a complement of a word (e.g., of a set of pixels or of bits).

Object. A component or a set of components of the same type within a digital image.

Online processing. Method of processing while being in direct communication with and under the direct control of the CPU (central processing unit) of the computer that executes the program.

Opening. A sequence of an erosion and a dilatation performed on a set (image) by a structuring element. First the image is eroded and then the eroded image is dilatated.

.OR. A logical operator that computes a word as the result of the union between two words.

Padding. One or more words placed at the beginning and end of each logical record, which stores one row of image data; one or more rows of data placed before the first row and after the last row of image data. Its purpose is to map a structuring element when some of its pixels are placed outside the edge of the image and its center pixel coincides in position with pixels at the edge. Generally, padding data consist of 0's.

Parallel computer. A computer that can process either the entire image or a large part of it simultaneously. For this its architecture consists of several computers working in a parallel fashion.

Pattern. A broad term here used to indicate either a binary image or part of it, or the information extracted from one or more images after processing.

Pattern recognition. Techniques which use sets of property values to characterize an image (Rosenfeld, A., and A. C. Kak, 1976, Digital Picture Processing, Academic Press, New York, pp. 404-405.) automatically.

PE. *See* Processing element.

Phase. The set of points (pixels) that includes all particles of a single type; the set of all particles of a single kind within an image.

Phase-labeled image. Image in which all components belonging to a same phase are identified by pixels with symbolic values corresponding to a given phase number.

Phase labeling. The process of producing a phase-labeled image.

Picture algebra. *See* Picture language.

Picture language. The result of attempts to create a computer language (like FORTRAN or PASCAL) in which to describe digital images (structure, composition, etc.); also called *picture algebra.* An example is the Golay alphabet (Golay, M. J. E., 1969, Hexagonal Parallel Pattern Transformations, *IEEE Trans. Comput.* **C-18:**733-740).

Picture element. *See* Pixel.

Picture processing. *See* Image processing.

Pipeline processing. Processing in which computations are structured in a pipeline fashion so that processing time is optimized for particular tasks. Neighborhood transformations can be very efficiently implemented on a pipeline processor.

Pipeline processor. A computer that processes data in a pipeline fashion.

Pixel. The element of a digitized image (array) called picture element, pixel, pel, or sometimes just point. A pixel is the value stored in a regular array of values (a digital image) in point-to-point correspondence with small areas in the original input material.

Point-to-point correspondence. The spatial relationship among pixels in a digital image and positions in the original input material.

Preprocessing. Operations and transformations commonly performed on an image preliminary to property measurements.

Processing element (PE). The logic and the memory associated with the cell when cellular logic is implemented in an array of computers or a parallel computer (or pipeline processor).

Profile. Object or silhouette of objects in a binary image. Profiles are the result of sectioning some material or of projections onto a plane.

Quantization. Process applied to the samples of optical digitization to obtain numbers having a discrete set of possible values (generally integer values). *See also* Digitization.

Random-access memory. *See* Read-write memory.

Raster. A regular spatial arrangement of pixels so that they occupy the corners of a rectangular grid (square raster) or of a triangular grid (hexagonal raster). When stored in computer memory, the pixels are implicitly assumed to occupy their position in such a raster, and are distributed from left to right and top to bottom in columns and rows.

Raster scanning. Optical digitization according to a regular spatial arrangement of pixel positions or raster.

Read-only memory. (ROM). Memory with direct-read access only.

Read-write memory. Memory with direct-read and direct-write access; also termed random-access memory (RAM).

Real time. Method of processing in which the computer responds to, and this response affects, the activity that is under the computer's control.

Refreshed-type display. Display device in which the refresh (redisplay) process consists of reading the information out of the memory and redisplaying it every 1/30 second.

Registration. Point-to-point correspondence between different images covering a same area, and of an image with the original input material.

Scanner. A mechanical or electronic instrument that measures the gray level tones from an original picture material and transforms them into digital values. Measurements are generally made in a regular raster fashion.

Scene analysis. Methods for deriving a useful descriptive relational structure from a given image or scene; also called *picture analysis.* Scene analysis is generally used only in connection with "three-dimensional" scenes in which perspective and occlusion by other objects play significant roles (Rosenfeld, A., and A. C. Kak, 1976, Digital Picture Processing, Academic Press, New York, p.440).

Scratch area. A portion of memory on disk that is reused during computations.

Sequential machine. A computer that executes instructions in sequence: information or data records are processed in the same order in which they happen.

Serial number. *See* Label.

Shifting operation. An operation or transformation that translates the bits within a word (or an array of words) either to the right or to the left. It is used together with logical operations to compute neighborhood transformations.

Skeletonization. The process of identifying the set of pixels within a set of objects in an image whose distances from object borders are locally maximum (i.e., no neighboring pixel has a greater distance from both sides of the borders or edges). If the pixels of the skeleton and the associated distance values are known, the objects can be reconstructed exactly. *See also* thinning.

Skeletons. A group of objects in a binary image; also called *medial axes* or *symmetric axes*. *See also* Skeletonization.

Software. The programs used by a computer system.

Square raster. Square array of pixels in which each point has either four immediately adjacent pixels at the same distance from it (up, down, left, and right) or eight immediately adjacent pixels (the eight pixels surrounding a pixel at the center of a 3×3 array) at two different distances (1 and $\sqrt{2}$) from it.

Star of the pores. The mean area seen directly from a point of the pores. In a digital image it is represented by the set of pixels seen directly, in the directions of the raster, by every pixel belonging to the pores.

Stereology. The field of study of the geometric properties in three dimensions from two-dimensional sections or projections through solid materials.

Storage display screen. A TV monitor on which image data are displayed and stored once in a stable manner, as opposed to a refreshed type of screen.

Strokes. Elementary parts of objects to be recognized.

Structuring element. A set of pixels that is swept across every pixel of an image whose 0-1 value is changed according to the degree of matching in its corresponding neighborhood. *See also* Neighborhood operator.

Synneusis. Process that produces a texture of rocks in which the individual crystals of a mineral species are collected to form groups or aggregates.

Template. An idealized prototype or a known object (e.g., a template of characters to be matched against pictures of printed pages, or a star pattern against the picture of the sky). It can be a piece of the digital image itself, or a piece of another image.

Template matching. A classical approach to the problem of locating an object in an image. An image is searched for an object by applying a template at each location in the image, and some measure of the extent to which the template matches (or does not match) the subimage at that location is computed to decide on the existence of the object at the given location by comparing this measure with a predetermined threshold.

Texture. Geometric aspects of component particles (of a rock or other materials or pictures), including size, shape, and arrangement.

Thinning. Processing to reduce any object in an image into a set of arcs and curves (and junctions) without changing the connectedness properties of the objects (four-, six-, or eight- adjacency). These arcs should consist of points (pixels) of the objects whose distances from the edges of the objects in the image are all approximately half the width of the respective objects. Many algorithms exist for thinning. *See also* Skeletonization.

Thresholding. Process to extract a two-valued (binary) image from a gray-level image.

Time-shared computer. A computer in which processing is shared (or, due to its fast response time, appears to be shared) by several users simultaneously.

Topology-preserving operation. An operation or transformation of an image that preserves the identity and relative position of the objects contained in it. Examples are thinning and labeling.

Transform. The image resulting from a transformation process.

Transformation. The process by which an image changes into another image, which is hoped to be more informative or, in any case, more useful.

Union. For any image subset A and B, the union $A \cup B$ is the set of points (pixel values) that are in either A or B, or both.

Universal set T (T0). The entire image set before a given sequence of transformations is performed. With it the transformation results are compared.

Window. A reserved area in a two-dimensional array that is dedicated to some special purpose or calculation.

Word. A group of bits placed together and treated as a unit in a single location in memory. Computer memory is logically organized in words; a word is one logical unit of information.

Word length. In computer terminology, the number of bits in the word of the computer. Different computers have, for example, 8, 16, 32, or 60 bits per word.

Index

237

About the Author

Andrea G. Fabbri held for many years the position of research scientist in the Geomathematics Section of the Geological Survey of Canada prior to being appointed senior research scientist with the National Research Council of Italy in 1983 at the Institute of Marine Geology in Bologna. He received his "Laurea di Dottore in Geologia" degree from the University of Bologna, Italy, and a Ph.D. in geology from the University of Ottawa, Canada. Before 1977, his main fields of activity were the quantification of geological variables and the preparation of computer-based geological data banks for statistical analysis as an aid to regional mineral resource evaluation. Since 1977, when he became a guest worker in the Electrical Engineering Division of the National Research Council of Canada, he has developed methods of image processing in geology for map digitization, geoscience data integration, thematic mapping, stereology, texture analysis, pattern recognition, and statistical analysis.

While on leave during 1973-74, Fabbri was head of the Section of Informatics at GEOTECNECO, a company of the E.N.I. Group, in Italy. During 1981-82, he held a visiting professorship at the University of Bologna, where he lectured on mathematical geology at the Institute of Mineralogy and Petrography.

Fabbri has authored and coauthored over 40 papers on subjects of quantitative geology. He is a member of the International Association on the Genesis of Ore Deposits, the Canadian Institute of Mining and Metallurgy, the International Association for Mathematical Geology, the International Society for Stereology, and the Pattern Recognition Society.

DATE DUE

OCT 7 '85			